Synthesis and Polymerization of Metal-Containing Monomers

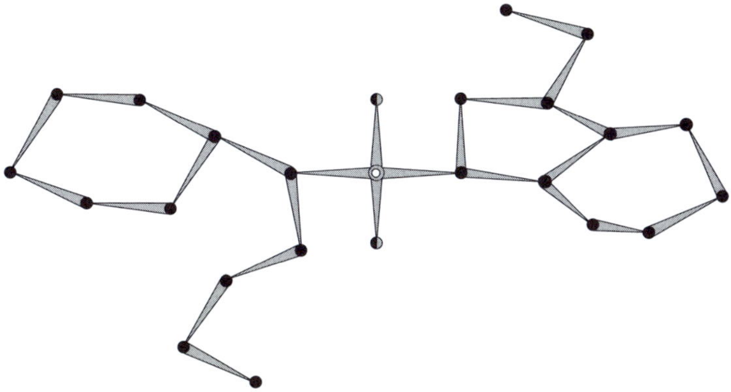

Anatoly D. Pomogailo, D.Sc.
Institute of Chemical Physics
Russian Academy of Sciences
Chernogolovka, Russia

Vladimir S. Savost'yanov, Ph.D.
Institute for Energy Problems of Chemical Physics
Russian Academy of Sciences
Chernogolovka, Russia

CRC Press
Boca Raton Ann Arbor London Tokyo

Library of Congress Cataloging-in-Publication Data

Pomogailo, A.D. (Anatolii Dmitrievich)
(Metallosoderzhashchie monomery i polimery na ikh osnove. English)
Synthesis and polymerization of metal-containing monomers/Anatoly D. Pomogailo and Vladimir S. Savost'yanov.
p. cm.
Includes bibliographical references and index.
ISBN 0-8493-2863-2
1. Organometallic polymers. I. Savost'yanov, Vladimir S. II. Title.
QD381.8.P65513 1994
547'.050459—dc20 93.49461
CIP

Direct all inquiries to CRC Press, Inc., 2000 Corporate Blvd., N.W., Boca Raton, Florida 33431.

International Standard Book Number 0-8493-2863-2
Library of Congress Card Number 93-49461
Printed in the United States of America 1 2 3 4 5 6 7 8 9 0
Printed on acid-free paper

CONTENTS

Table 2 is reprinted from Hatada, K., Nakanishi, H., Ute, K., and Katayama, M., *Polymer Journal,* 18, 58, 1986. With permission.

Figure 5 is reprinted from Doppert, K., Sanchez-Delgano, R., Klein, H.-D., and Thewalt, U., *Journal of Organometallic Chemistry*, 233, 205, 1982. With permission.

Figure 28 is reprinted from Grissier, J. C., Levesque, G., and Patin, A., *Polymer Bulletin*, 8, 55, 1982. With permission.

PREFACE

This monograph is devoted to a relatively new branch of chemistry that has emerged in the last two or three decades at the junction where the organometallic, coordination, and high molecular branches of chemistry intersect. Because of the extensive efforts of researchers from different countries, the syntheses and polymerization conversions of organometallic monomers have been developed to the extent that it has become possible to speak about an entirely new interdisciplinary science. It possesses all of the features typical of an independent branch of chemistry (such as unique objects of studies) that have found wide applications and are distinguished by specific structural properties, powerful methods for polymerization conversions, and studies of such monomers and sensitive techniques to investigate the structure of the appearing products, to name a few. Finally, there exists a vast field where metal-containing polymers are used to advantage due to the multifaceted properties of metals enriched with the polymeric nature of the matrix. Upon polymerization, metal-containing monomers give rise to composite materials having new physicochemical and functional properties, catalytically and biologically active macrocomplexes, etc.

Therefore it is only natural that along with general conferences and symposia dealing with the chemistry of high molecular, coordination, and organoelement compounds, as well as catalysis (where much attention is being paid to metal-containing monomers and polymers), there have been a number of special international gatherings on the problems that form the subject of this book. Also, they have been treated in detail by C.U. Pittman, Jr., C.E. Carraher, Jr., J. Sheats, Z. Wojtczak, C.R. Simíonescu, and other researchers. In Russia the same subject has been extensively studied, primarily by V.V. Korshak and V.A. Kabanov.

For these reasons the authors of this volume consider the generalization of numerous experimental and theoretical works as an urgent and a timely issue. First of all, this is associated with a need for coordinating the efforts of specialists whose interests lie in the various fields of chemistry. Moreover, a well-planned approach to the solution of problems of synthesizing metal-containing monomers and also their polymerization and copolymerization is capable of giving a tool for designing a macromolecule at the stage of creating metal-containing polymer systems and programming the desired properties for them.

Further progress in the area under discussion may well be visualized as lying on a path of developing economically advantageous and ecologically clean processes for producing the most important metal-containing monomers and the associated polymers. Moreover, this field is very promising from the theoretical point of view. As a matter of fact, polymerization conversions of metal-containing monomers provide a unique possibility for inserting practically any metal—in all its valent states, ligand environments, and polyhedron configurations—in the polymer chain. At present, the major theoretical considerations concerning structure of metal-containing monomers, their polymerization, and structure of the products are most often based on the logical analysis of experimental data and general scientific statements. However, there is every reason to believe that in the near future a firm research foundation will be established for this branch of chemistry. This conviction is supported by current developments in the field of metal-containing monomers and associated polymers.

The authors wish to express their gratitude to Mr. A.Z. Aravsky whose creative approach to the translation of this book has made it accessible to the audience. Our thanks are also due to our colleagues from various research establishments, who have made many valuable suggestions and comments to improve the content of the book.

It is our hope that this volume will not be lost among the publications devoted to the development of this interesting and promising domain of science but will attract more attention to its problems.

A.D. Pomogailo
V.S. Savost'yanov

Chernogolovka
August 1993

AUTHORS

Anatolii D. Pomogailo, D.Sc. (Chem.), is Professor of Polymer Chemistry at the Institute of Chemical Physics, Russian Academy of Sciences, in Chernogolovka. Born in 1939 in Kiev Region (Ukraine), he graduated from Odessa State University (Ukraine), Department of Physics and Chemistry of Polymers. He received his Ph.D. in 1970 from the Institute of Chemical Sciences (Kazakh Academy of Sciences). The title of his Ph.D. thesis was "Kinetics and mechanism of catalytic (co)polymerization of ethylene by modified systems on the base of complexes of high valency titanium and vanadium halids". Since 1972, Dr. Pomogailo is working at the Institute of Chemical Physics, Russian Academy of Sciences. He received his Doctorate in Chemistry in 1981 from the Institute of Chemical Physics. The title of his thesis was "Immobilization of Metal Complexes by Macromolecular Supports and Catalytic Properties of Such Systems in Polymerization Reactions". He received the title of Professor in 1991. Since 1982 he has been the head of a group involved with metal-containing polymers that was transformed in the laboratory in 1993.

The scope of his scientific interest includes: immobilized metal complexes, catalysis by macromolecular complexes, (co)polymerization of alfa-olefines, and metal-containing monomers and polymers. Dr. Pomogailo is the author of over 250 articles and about 100 patents. He has also authored four monographs: "Polymer-Immobilized Metal Complex Catalysts", Moscow, Nauka, 1988, 303 pp, in Russian; "Catalysis by Immobilized Complexes", Moscow, Nauka, 1991, 448 pp, in Russian; "Metal-Containing Monomers and their Polymers", Moscow, Khimiya, 1988, 384 pp, in Russian, with V.S. Savost'yanov; and "Macromolecular Metal Chelates", Moscow, Khimiya, 1991, 304 pp, in Russian, with I.E. Uflyand.

Vladimir S. Savost'yanov, Ph.D. (Chem.), is a researcher at the Institute for Energy Problems of Chemical Physics, Russian Academy of Sciences. Born in 1954 in the Moscow Region (Russia) he received his M.D. in 1977 from the Moscow Chemical Technological Institute and a Ph.D. in 1987 from the Institute of Chemical Physics, Russian Academy of Sciences.

Dr. Savost'yanov has been working in the field of metal-containing monomer polymerization. Originally he was interested in the radiation-induced graft polymerization of metal-containing monomers (metal acrylates and metal salts acrylamide complexes) onto polyethylene.

Dr. Savost'yanov's main interest at present is the investigation of frontal and spontaneous polymerization of metal nitrates acrylamide complexes.

LIST OF ABBREVIATIONS

AA	Allylamine
AAAm	*N-tert*-amyl acrylamide
AAc	Acrylic acid
AAl	Allyl alcohol
AAm	Acrylamide
AIBN	Azobisisobutyronitrile
AN	Acrylonitrile
BAAm	*N-tert*-butyl acrylamide
BP	Benzoyl peroxide
BVSO	Butylvinyl sulfoxide
CN	Coordination number
D	Exposure dose
Dipy	Dipyridyl
DMFA	Dimethyl formamide
DMPA	Dimethylolpropionic acid
DMSO	Dimethyl sulfoxide
DSC	Differential scanning calorimetry
DTA	Differential thermal analysis
DTGA	Differential thermogravimetric analysis
e	Polar factor
E_a	Activation energy
EMP	Ethylene glycol methacrylate-phthalate
ESCA	Electron spectroscopy for chemical analysis
ESR	Electron spin resonance
G_{-M}	Radiation yield
HAAm	N-tert-hexyl acrylamide
HMPA	Hexamethyl phosphoramide
HP	Hydroperoxide
I	Initiator, exposure dose rate
IR	Infrared
K_D	Diffusion coefficient
K_{diss}	Constant of dissociation
k_i	Constant of initiation rate
k_p	Constant of chain propagation rate
k_t	Constant of chain termination rate
L	Ligand
M	Monomer
MA	Methylacrylate
MAA	Methacrylic acid
MAn	Maleic anhydride
MAN	Methacrylonitrile
MCM	Metal-containing monomer
MM	Molecular mass
MMA	Methyl methacrylate
M_n	Number-average molecular mass
M_w	Weight-average molecular mass
MVP	2-Methyl-5-vinylpyridine
4-M-2VP	4-Methyl-2-vinylpyridine
MX_n	Metal compound
NMR	Nuclear magnetic resonance

P_n	Degree of polymerization
PAA	Polyacrylic acid
PAAm	Polyacrylamide
PE	Polyethylene
PETP	Poly(ethylene terephthalate)
Phen	1,10-Phenanthroline
PMAA	Polymethacrylic acid
PMMA	Polymethyl methacrylate
PMR	Proton magnetic resonance
PP	Polypropylene
PTFE	Polytetrafluoroethylene
PVC	Polyvinyl chloride
Q	Resonance stabilization parameter
r	Relative reactivity parameter
TBTA	Tributyltin acrylate
TBTM	Tributyltin methacrylate
T_d	Destruction temperature
T_f	Flow temperature
T_g	Vitrification temperature
T_m	Melting point
T_v	Vitrification temperature
THF	Tetrahydrofuran
Tpy	2,2, 2´´-Tripyridyl
UV	Ultraviolet
VA	Vinyl acetate
VBI	1-Vinylbenzimidazole
VBT	1-Vinylbenzotriazole
VIA	1-Vinylimidazole
VMIA	1-Vinyl-2-methylimidazole
VP	Vinylpyridine
2-VP	2-Vinylpyridine
4-VP	4-Vinylpyridine
VPr	*N*-Vinylpyrrolidone
w	Polymerization rate
w_{fr}	Front propagation rate
w_i	Initiation rate
w_0	Initial polymerization rate
w_p	Chain propagation rate
ν	Bending mode frequency
η	Inherent viscosity
λ	Wavelength
μ_B	Bohr magneton
μ_{ef}	Effective magnetic moment
δ	Stretching mode frequency

INTRODUCTION

Production of materials that combine the properties of metals and polymers has long been a matter of interest to researchers. There are a large variety of techniques to obtain such combinations, from preparing mixed polymer compositions with fine-powdered metals to the introduction of metal ions into preformed macromolecules using polymer-analogous methods.[1] Polycondensation of organometallic compounds is also used to advantage to obtain such products. As this takes place, the metal ions can be found in both the main and the side chains. Interestingly, all these materials are covered by the same broad term, namely, "metallopolymers".

However, this problem can be treated differently. According to the high molecular compound chemistry, metallopolymers are expected to be preceded by metal-containing monomers (MCMs). They can be regarded as compounds composed of a multiple bond (bonds) with its (their) ability to open and a chemically bonded metal-containing group. It should be noted that in the overwhelming majority of cases multiple MCM bonds are of the vinyl type, although there are rare examples of monomers containing allyl, ethynyl, and vinylethynyl groups.

Until very recently the literature contained no MCM classification of its own: MCMs were treated either as part of conventional monomers—based on the type of the appearing chain, nature of the multiple bond, types of substituents, etc.—or as belonging to organometallic compounds. Due to the fact that MCMs represent a special type of organoelement compounds, it is believed that their classification primarily should include the type of bond between the metal and the organic part of the molecule.[2, 3] According to this principle, MCMs can be divided into the following major classes: covalent (I), ionic (II), donor-acceptor (III), chelate (IV), and π (V) types:

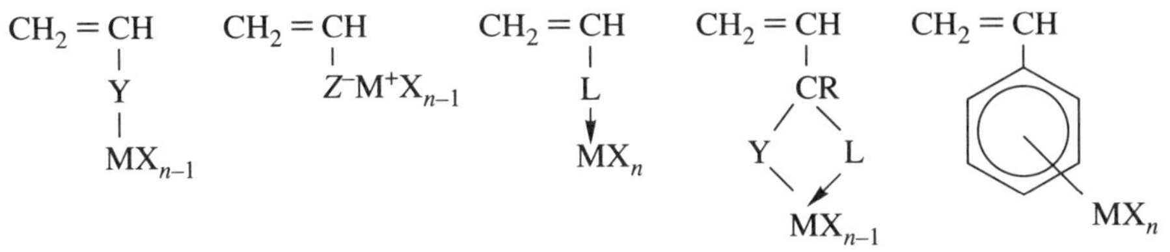

(I) (II) (III) (IV) (V)

where *M* is the metal; *n* is its valence; *X*, *Y*, and *Z* are the corresponding functional groups.

Moreover, along with monometallic single-atom complexes, MCMs can include heterometallic and cluster-type compounds for which the chemistry is at the early stage of development.

Covalent-type MCMs are responsible for a relatively small fraction. Primarily this pertains to true organometallic "metal-carbon" bond monomers, and a somewhat larger group incorporates "metal-oxygen" MCMs. Ionic-type monomers belong to the most widely spread group typical of both nontransition and transition metals. Donor-acceptor-type monomers, when the π-electrons of the multiple bond do not take part in complexing reactions, do not essentially differ from conventional (*nV*-type) complexes. They are most characteristic of transition metal halides. Chelate-type monomers represent a combined group of MCMs. They appear as a result of a covalent (ionic)- and *nV*-bonding. It was not until the last decade or so that researchers began studying these MCMs. While the functional groups in heteroatoms of the MCM ligands that give rise to donor-acceptor-type monomers provide lone electron pairs to form coordination bonds, in the event of π-type MCMs a π-electron system is involved in their formation. π-MCMs are characteristic exclusively of transition metals, predominantly of groups VIA, VIIA, and VIII of the Periodic Table.

Among the MCMs mentioned above, the greatest attention has been paid to π-type monomers. The syntheses, structure, polymerization, and copolymerization of these monomers (metallocene, cyclopentadienyl, styrene, and others) have been reviewed in excellent publications by U.S. authors.[4–7] For this reason as well as for consideration of space, this book does not describe these monomers.

The limited volume of this book also has prevented the authors from discussing polymerization conversions of "potential" MCMs with multiple bonds. The authors restrict themselves to MCMs with known polymerization properties. In addition, the same reason is responsible for the fact that the authors do not provide specific techniques suitable for performing polymerization conversions even with respect to the most important representatives of this class of monomers, as has been done in the enlarged Russian edition of the book.[3]

The order of presenting the material follows the above classification with respect to both MCMs and associated polymers. Within each class the material is presented in accordance with the position of a given metal in the Periodic Table.

It was our intent to give a well-balanced account of different viewpoints of the problems under discussion. Nevertheless, it is understood that in some cases the authors give preference to the opinions that reflect their own position as researchers.

Chapter 1

Production and Polymerization Conversions of Covalent-Type MCMs

Covalent-type metal containing monomers (MCMs) can include both true organometallic monomers containing an M—C bond and some of the M—O monomers. Knowledge about this class of compounds is rather limited. However, in the last few years the situation in this field has changed; this fact is corroborated by the appearance of new original works concerning both the synthesis and the polymerization conversions of these MCMs.

I. SYNTHESIS OF COVALENT-TYPE MCMs

It should be noted that monomers incorporating M—C and M—O bonds differ greatly in a large number of properties and the ways they can be synthesized, specifically the syntheses of true organometallic monomers.

A. SYNTHESIS OF TRUE ORGANOMETALLIC MONOMERS

These MCMs contain vinyl, allyl, styrene, and other groups directly associated with a metal. Extensive studies of their syntheses and properties date back to the development of preparative productions of organomagnesium compounds containing the corresponding radical (primarily a vinyl one, Normant's reagent).[8–10] The most convenient way for producing this reagent is via a reaction of vinyl halide with magnesium tetrahydrofuran (THF). The vinylmagnesium halides resulting from this reaction are interesting not by themselves but rather as intermediates used to synthesize unsaturated derivatives of some other metals.

Unsaturated organomagnesium compounds are rather unstable and sensitive to atmospheric oxygen, moisture, etc. Arylvinyl Mg(II)-containing species are more stable.[11] One of the best studied representatives of these MCMs, *p*- and *m*-vinylbenzenemagnesium halides:

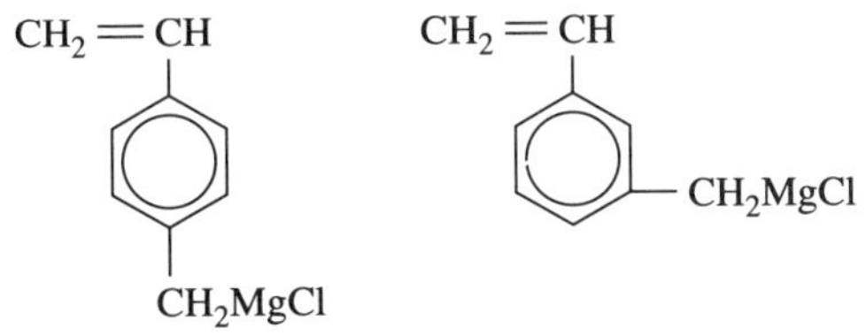

were obtained through the interaction of magnesium with vinylbenzyl chloride in ether.

Among vinyl derivatives the greatest recognition for polymerization conversions has been gained, perhaps by metal compounds of Group IV of the Periodic Table. Vinyl compounds of Ge(IV) are synthesized by reacting the Normant reagent with alkyl- and arylgermanium in THF:[10,12]

$$R_{4-n}GeCl_4 + nCH_2 = CHMgX \rightarrow (CH_2 = CH)_n GeR_{4-n} + nMgXCl.$$

Vinyl derivatives of tin are obtained in the same way.[10] This metal is specific in that in this case the synthesis can be conducted, along with halides and with organotin oxides:[13]

$$(C_4H_9)_3SnOSn(C_4H_9)_3 + 2CH_2 = CHMgCl \rightarrow$$
$$\rightarrow 2(C_4H_9)_3SnCH = CH_2 + MgO + MgCl_2 .$$

Halogen-containing vinyl derivatives of tin have been conveniently synthesized via a redistribution reaction involving vinyl compounds of this metal and Sn(IV) tetrachloride.[14] In addition, this gives rise to such individual products as $(CH_2{=}CH)_3SnCl$ and $(CH_2{=}CH)_2SnCl_2$.

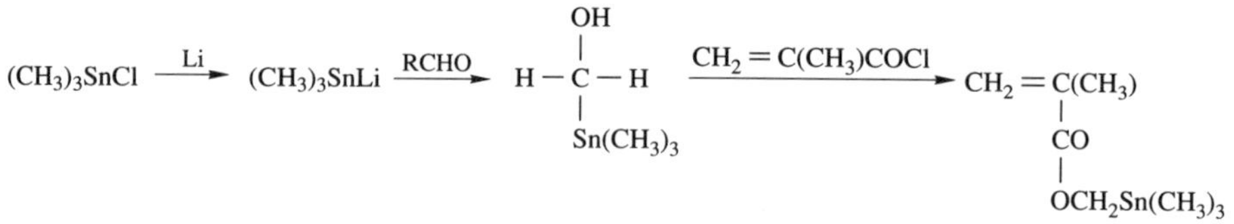

Also, mention should be made of a recently synthesized new group of organotin monomers, namely, 1-(trimethylstannyl) alkyl acrylates:[15]

Vinyl Tl(III) derivatives have been effectively synthesized using the Normant reagent:[16]

$$CH_2 = CH\,MgX + TlX_3 \rightarrow Tl(CH = CH_2)_2\,X$$

where X = Cl or Br.

The resulting products can then be used for producing the corresponding Hg(II) and Sn(IV) derivatives:

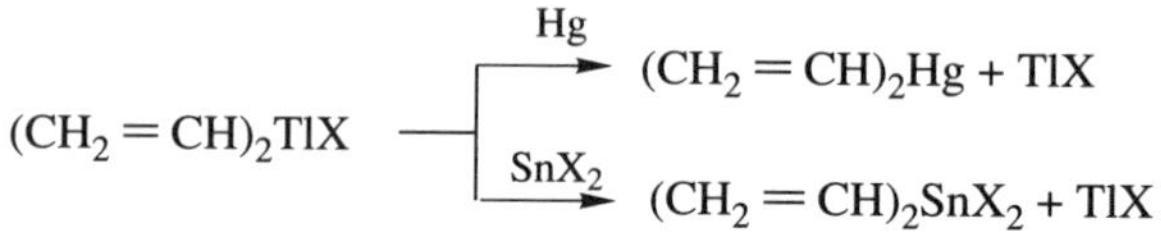

Divinylmercury is in turn a good vinylating agent for the synthesis of the corresponding Al(III) derivatives.[17]

Normant's reagent also can be applied to the synthesis of vinyl Pb(IV) derivatives:[18]

$$4CH_2 = CHMgX + 2PbY_2 \rightarrow Pb(CH = CH_2)_4 + 2Pb + 4MgXY,$$

where X = Cl or Br, and Y = Cl, I, or CH_3COO.

Among covalent-type MCMs there is a very interesting group of organometallic styrene derivatives based on germanium, mercury, tin, and lead. Their synthesis has been well studied elsewhere:[19–21]

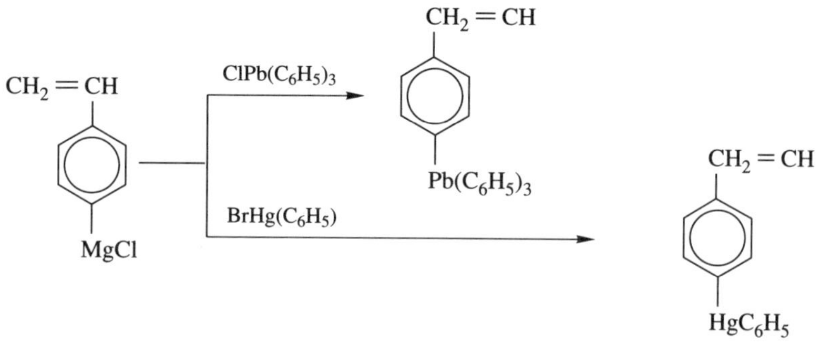

α-Methylstyrene derivatives can be produced using both organolithium synthesis:[22]

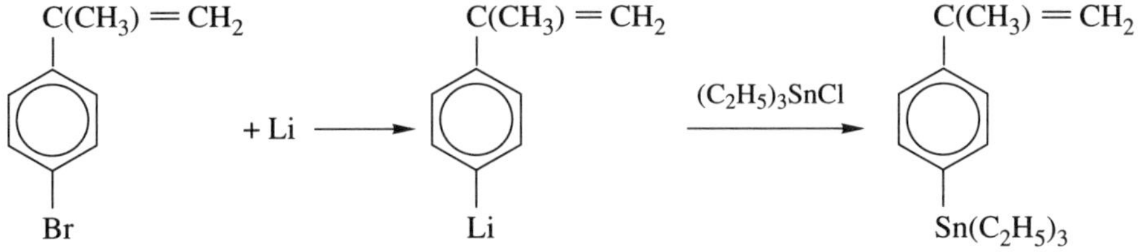

or the Normant reagent:[23]

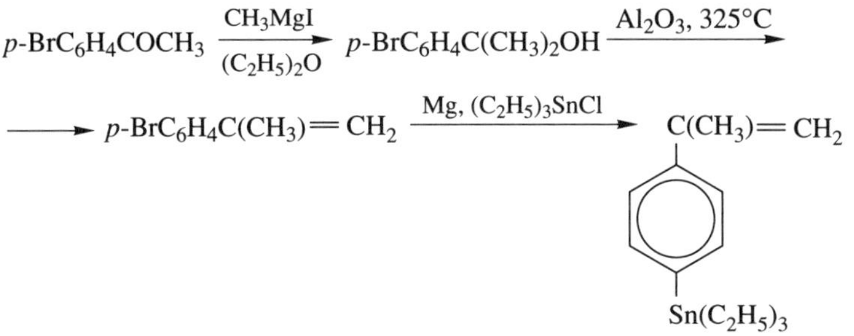

Similar methods are used to prepare styrene derivatives of other metals.

Among transition metal-containing acetylene derivatives that are few in number one can mention only bis(1,3-pentadiinyl)mercury[24] $H_3C{-}C{\equiv}C{-}C{\equiv}C{-}Hg{-}C{\equiv}C{-}C{\equiv}C{-}CH_3$ since the polymerization properties of other representatives of these compounds have not been studied.

Concerning transition metal-containing organic compounds, the corresponding vinyl derivatives are usually very unstable, which complicates their application to polymerization processes. At the same time, it has been possible to obtain the following stable Pd(II) and Pt(II) compounds:[25]

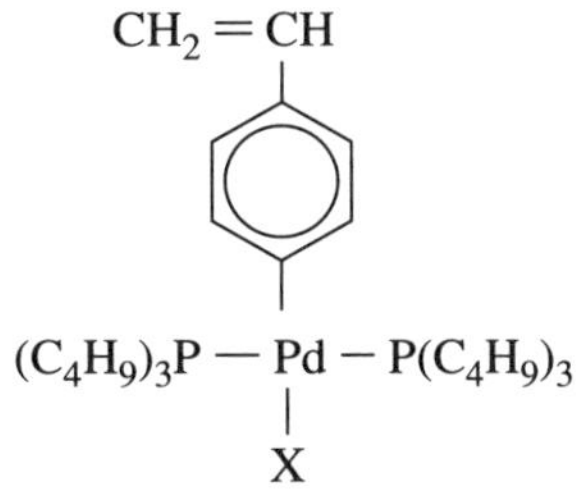

and

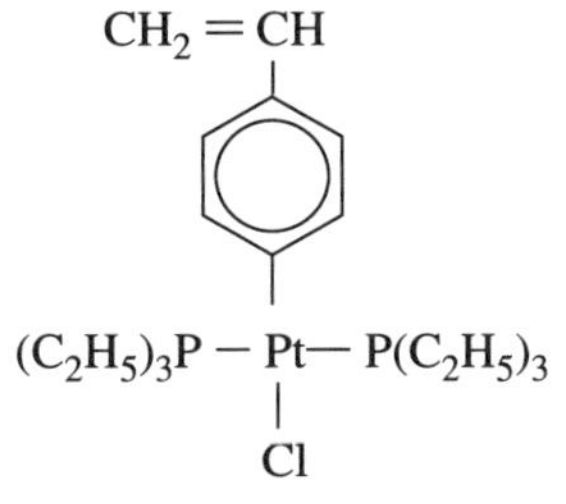

where X = Cl, Br, CN, or C_6H_5.

Such MCMs are synthesized in THF starting with vinylphenylmagnesium bromide and *trans*-dichlorobis(tributylphosphino)palladium(II) with subsequent splitting of the distyryl derivative to a monostyryl one (in ether at 70°C with excess HCl):

$$[P(C_4H_9)_3]_2PdCl_2 + 2CH_2{=}CHC_6H_4Br \longrightarrow$$

$$trans\text{–}[P(C_4H_9)_3]_2Pd(C_6H_4CH{=}CH_2)_2 \xrightarrow{\text{HCl, ether, –30°C}}$$

$$trans\text{–}[P(C_4H_9)_3]_2Pd(C_6H_4CH{=}CH_2)Cl \xrightarrow{\text{acetone, LiBr, NaCN}}$$

$$trans\text{–}[P(C_4H_9)_3]_2Pd(C_6H_4CH{=}CH_2)X \xrightarrow{C_6H_5MgBr}$$

$$trans\text{–}[P(C_4H_9)_3]_2Pd(C_6H_4CH{=}CH_2)C_6H_5$$

A platinum-containing monomer is obtained in a similar manner, except that the second vinyl group is split off at room temperature with an equimolar amount of HCl.

B. SYNTHESIS OF COVALENT-TYPE MCMs CONTAINING AN M—O BOND

Monomers with a covalent M—O bond constitute a fairly narrow class. As a matter of fact, they include almost exclusively derivatives of unsaturated alcohols such as vinyl (which is stable in its compounds only), allyl, and propargyl. In this case the selection of metals is also very limited: the only published data are available on MCMs based on Mg(II), Al(III), Ti(IV), and V(V). We are aware of the following Al(III)-containing monomers:[26]

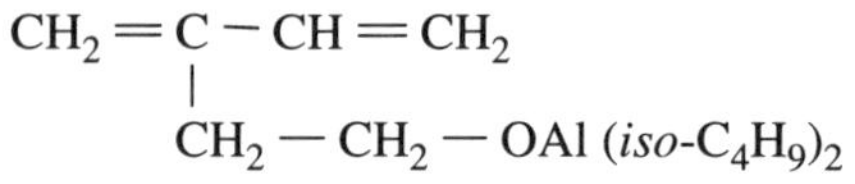

and

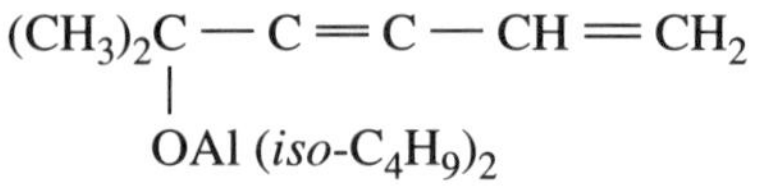

Thus, a reaction of dibutylmagnesium with ethylene glycol methacrylate gave the corresponding unsaturated alcoholate:

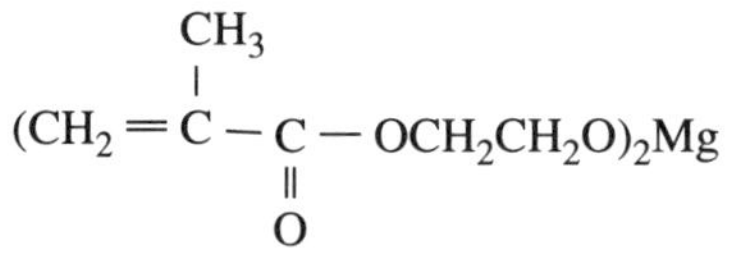

Monomers based on Ti(IV) are much more numerous. Such MCMs are usually, synthesized by reacting an unsaturated alcohol with a metal compound via a direct interaction of the alcohol with $TiCl_4$, the interaction of alcohol (alcoholate) with alkoxytitanium halide, substitution of the saturated alkoxy groups in alkyl titananes by the unsaturated ones, etc.[27]. In the first case the reliance is placed on a high reactivity of the Ti—Cl bond, sometimes using a stoichiometric mixture of alcohols one of which is unsaturated. An ammonium-pyridine modification of this method consists of a reaction

Table 1 **Syntheses and some properties of unsaturated Ti(IV) alcoholates**

MCM	Method	T_b, °C/Pa	Ref.
$Cl_3TiOCH_2CH{=}CH_2$	$TiCl_4 + CH_2{=}CHCH_2OH + NH_3 + Py$		35
$ClTi(OCH_2CH{=}CH_2)(C_5H_5)_2$	$(C_5H_5)_2TiCl_2 + CH_2{=}CHCH_2OH + NH_3$		24
$(C_3H_7O)_3TiOCH_2CH{=}CH_2$	$TiCl_4 + CH_2{=}CHCH_2OH + 3C_3H_7OH + NH_3 + Py$	222–223/400	35
$(C_4H_9O)_3TiOCH_2CH_2OCOC(CH_3){=}CH_2$	$Ti(OC_4H_9)_4 + HOCH_2CH_2OCOC(CH_3){=}CH_2$	153–159/130	33
$(C_4H_9O)_3TiOCH_2C{=}CHCH{=}CH$ (ring closed through O)	$Ti(OC_4H_9)_4$ + furfuryl alcohol	154–159/130	33
$(C_4H_9O)_3TiOC(CH_3)_2C{\equiv}C{-}CH{=}CH_2$	$Ti(OC_4H_9)_4 + HOC(CH_3)_2C{\equiv}C{-}CH{=}CH_2$	152–155/90–130	33
$(iso\text{-}C_3H_7O)_2Ti(OCH{=}CH_2)_2$	$Ti(O\text{-}iso\text{-}C_3H_7)_4 + 2CH_3CHO$		34
$Ti(OCH{=}CHC_5H_{11})_4$	$Ti(O\text{-}iso\text{-}C_3H_7)_4 + 4C_6H_{13}CHO$	70	34
$Ti(OCH{=}CHC_6H_5)_4$	$Ti(O\text{-}iso\text{-}C_3H_7)_4 + 4C_6H_5CH_2CHO$	92–94/70	34
$(iso\text{-}C_3H_7O)_2Ti(OC(CH_3)CH{=}C(CH_3)_2)_2$	$Ti(O\text{-}iso\text{-}C_3H_7)_4 + 2CH_3COCH{=}C(CH_3)_2$		34

in the presence of pyridine and/or NH_3 which binds the liberated HCl. This permits high yields and avoids side reactions such as those occurring in the event of allyl alcohol. The principal disadvantage of this method lies in the formation of a large number of fine precipitates. To eliminate this disadvantage, it has been proposed to perform syntheses in a system composed of two immiscible liquid phases; one is a solvent for the appearing precipitate, and the other is the same for the desired product (e.g., in a dimethyl formamide [DMFA]-hexane mixture, or in the presence of liquid ammonium).[28,29] It seems convenient to produce tetraallyl orthotitanate from $TiCl_4$ and allyl alcohol in a toluene-NH_4NO_3–NH_4 system.[30]

Similar approaches can be applied to the production of MCMs from halogen-containing cyclopentadienyl Ti(IV) derivatives:[31]

$$(C_5H_5)_2\,TiCl_2 + CH_2 = CHCH_2OH + NH_3 \rightarrow (C_5H_5)_2\,Ti(OCH_2CH = CH_2)Cl.$$

However, attempts to obtain bis(allyloxy)titanocene from 1 mol of $(C_5H_5)_2TCl_2$ and 2 mol of allyl alcohol result in the destruction of the titanocene structure.

Unsaturated alkoxy Ti(IV) derivatives are most often synthesized by transesterification. This method is applicable to those cases when the substitution is conducted using an alcohol with a boiling point that is higher than that of the alcohol to be substituted.[32] It is equally suitable for producing mixed alkyl orthotitanates:

$$Ti(OR)_4 + nR'OH \rightleftarrows Ti(OR)_n(OR')_{4-n} + nROH.$$

The released alcohol can be removed from the reaction mixture either directly during the synthesis or as a solvent (e.g., benzene-containing azeotrope). This method has been used to obtain a large series of monosubstituted alkoxy derivatives $(C_4H_9O)_3Ti(OR')$, where R′ is the residue containing reactive double, conjugated double, triple and conjugated double, and triple bonds; R′OH is the monomethylmethacrylic ethylene glycolate, furfuryl and propargyl alcohols, dimethylvinylethynylcarbinol, and vinylethynylpiperidol.[33] The addition of a large amount of OR′ groups to the molecule can lead to a spontaneous oligomerization of the resulting MCMs.

Note also such a specific method of MCM synthesis as the interaction of $Ti(OR)_4$ with enolizing aldehydes and ketones. It has been useful in producing tetrastyryloxy-, dihexadienyloxy-, and divinyloxydiisopropoxy titanates:[34]

$$Ti(OR)_4 + nR''_2CHCR'O \rightleftarrows Ti(OR)_{4+n}(OCR' = CR'')_n + nROH$$

where R′ and R′′ = H, alkyl, or aryl.

The methods for synthesizing Ti(IV)-containing covalent-type monomers are summarized in Table 1.

To synthesize V(V)-containing monomers, transesterification reactions can be used to advantage.[36] It has been possible to obtain and to identify MCMs containing one, two, or three unsaturated groups of enthylene glycol monomethylmethacrylate $OV(OC_3H_7\text{—}iso)_{3-n}(OCH_2\underset{\underset{O}{\|}}{C}C(CH_3){=}CH_2)_n$, where $n = 1, 2, 3$. When other unsaturated alcohols such as propargyl, allyl, and dimethylvinylethynyl are used in the reaction, one can observe the formation of mixed MCMs only which in addition to that are rather unstable. Especially unstable are V(V)-containing monomers based on allyl and propargyl alcohols, which (even when stored for a short time [1 to 1.5 h] after being produced) decompose to give black oil.

II. HOMOPOLYMERIZATION OF COVALENT-TYPE MCMs

A. HOMOPOLYMERIZATION OF TRUE ORGANOMETALLIC MONOMERS

The literature contains no worthwhile systematic data on homopolymerization of the MCMs under discussion (the same refers to copolymerization, however). This can be attributed to problems associated with the synthesis of these species, their low stability, and side reactions frequently involved in the polymerization process. Thus, in the case of monomers for which the unsaturated group is directly linked with the metal, the metal hydride can split off:[4]

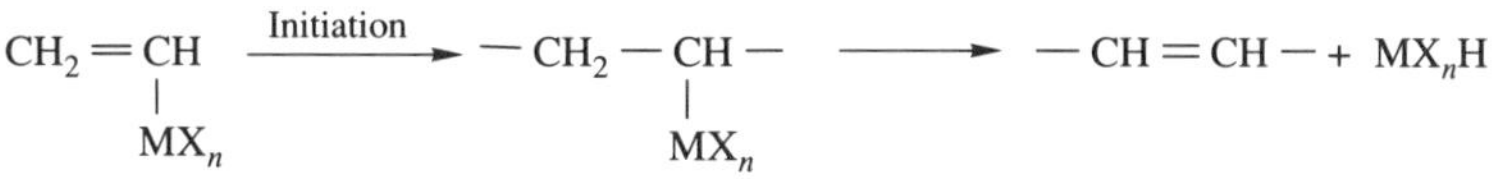

The probability of this process sharply decreases as the multiple bond and the metal are separated, as is the case of allyl, styryl, and (meth)acrylic monomers.

It should be noted in this context that the M—C bond is much more reactive than M—O, M—N, M—Cl, M—Br, and others. In addition, the strength of the M—C bond in transition metal compounds is very much the same as that of a similar bond in the major metal groups. However, with transition metals at the same excitation energies one can observe a more significant bond distortion.

Among the data relating to the homopolymerization of these MCMs it would be reasonable to stress those studies that are concerned with polymerization conversions of *p*- and *m*-vinylbenzylmagnesium chlorides.[11] The data for MCMs are consistent in diethyl ether at room temperature; however, as this solvent is replaced by THF, these monomers undergo spontaneous polymerization. At 0°C the process is accompanied by precipitation of a white solid polymer (Table 2). The rate of the process shows a drastic increase as the temperature is raised to 60°C, but in this case an insoluble polymer is formed. Interestingly, the resulting Grignard polymeric reagent contains a multiple bond at the end of the

Table 2 **Homopolymerization of vinylbenzylmagnesium chlorides[11]**

MCM[a]	Solvent	T °C	Time (d)	Yield %	$\overline{M}_n$	$\overline{M}_w/\overline{M}_n$
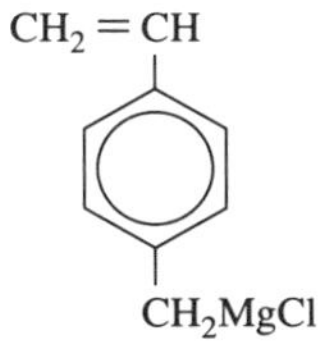	$(C_2H_5)_2O$	0	7	0		
	THF	0	3	2.5	7800	3/3
	THF	60	1	11[b]		
	HMPA/toluene	0	3	78	4900	3/7
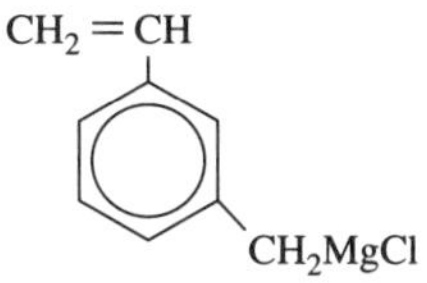	THF	0	3	11	3300	1/8
	THF	60	1	33[b]		
	Toluene	0	3	0		

[a][MCM] = 1 mol/l.

[b]Polymer insoluble in standard solvents.

chain. The polymerization process is likely to take place along the lines of organometal synthesis with the monomer being inserted at the M—C bond:

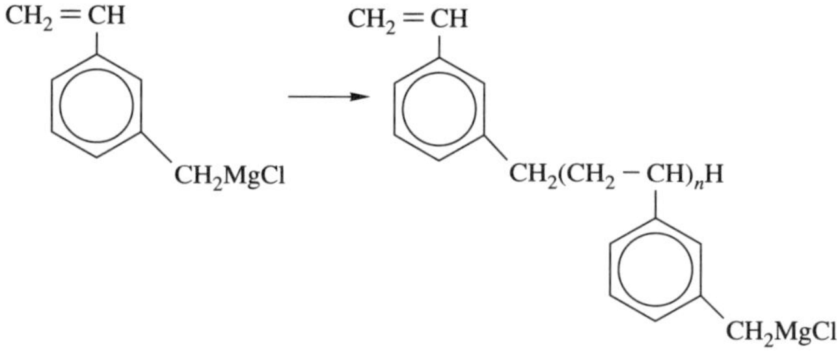

The polymer obtained in the presence of such a solvent as hexamethyl phosphoramide (HMPA) is very branched, which is linked with an enhanced activity of the residual benzylmagnesium chloride groups, this activity being sufficient for MCM polymerization:

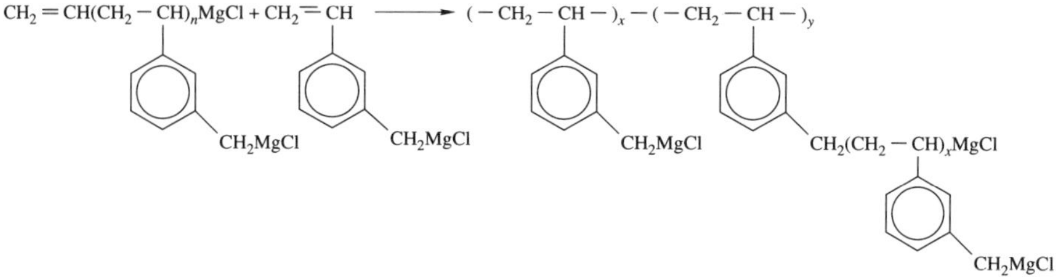

The reactivity of the *m*-isomer was higher than that of the *p*-isomer.

p-Styrylmagnesium chloride was homopolymerized in a 30% yield during the devitrification of the irradiated and prefrozen MCM solutions in THF (for more information on the method used for this purpose see Chapter 2).[53]

Homopolymerization of some vinyl, allyl, and styryl monomers based on Ge(IV), Sn(IV), and Pb(IV) has been studied fairly well. The reactivity of trimethylvinyl and triethylallyl MCMs was compared with the data relating to the polymerization of the corresponding carbon- and silicon-containing monomers at high pressures (6000 atm) using *tert*-butyl peroxide as the initiator at 120°C.[37] The series of activities obtained, Si(IV) > C(IV) > Ge(IV) > Sn(IV), is valid for both vinyl and allyl derivatives. The polymerization ability of these MCMs was rather low: degree of polymerization (P_n) does not exceed 5 and polymer yields are at a level of 30 to 35%.

It should be noted that the majority of vinyl- and allyl germanium compounds are not polymerizable when use is made of radical initiators and when under normal pressure.[12, 38] In the absence of positive pressure, homopolymerization can be observed for those monomers that contain three or four unsaturated groups.[38,39] The same is true of tin and lead derivatives. The methylallyl derivative $(CH_3)_3GeCH_2C(CH_3){=}CH_2$ is not polymerizable by the radical mechanism and under high-pressure conditions.

In this context, it would be very interesting to attempt the use of other initiators. Thus, the initiated polymerization of trimethylvinylgermanium and stannane with C_4H_9Li in heptane at 20°C gives polymers with molecular mass (MM) = 6 to $9{\cdot}10^3$.[40] Dimethyldiallylgermanium gives polymers in the presence of Ziegler-Natta catalysts, e.g., $TiCl_4$—$Al(C_2H_5)_3$.[41] The same refers to vinylgermanium $CH_2{=}CH$—GeH_3 readily polymerizable to a high molecular product. On the other hand, $CH_2{=}CHPb(CH_3)_3$ polymerization on a $TiCl_4$-AlR_3 system meets some difficulties. The only polymers formed are those with a low MM.[42]

The susceptibility of other types of MCMs, namely, organometallic styrenes, R_3M—C_6H_4—$CH{=}CH_2$, to radical copolymerization diminishes in the order Sn(IV) > Pb(IV) >

C(IV).[19,21] According to other data, the reactivity of such MCMs is different. It matches the activity changes in the polymerization of vinyl and allyl compounds of Pb(IV), C(IV), Si(IV), Ge(IV), and Sn(IV) in the same order.[43] Note an enhanced reactivity of a lead-containing styrene, which is attributed to a parallel homolytic breakage of the Pb—C bond to promote the polymerization process and to give rise to three-dimensional structures:

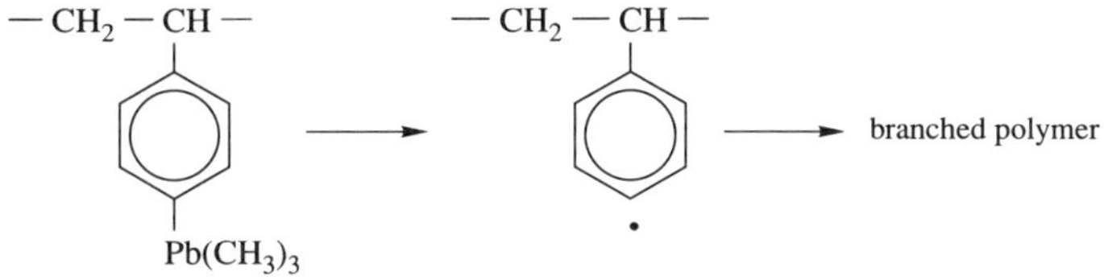

Generally, one can observe a decrease in the MCM polymerization ability with the atomic mass of the Group IV elements, which can be explained by the appearance (starting with silicon) of a conjugated MCM-vacant metal *d*-orbital bond (*p*-*d* conjugations). This results in a sharp drop of the electronic density and, accordingly, in the double bond passivation.

Also, the polymerization MCM properties are heavily affected by the type of substituents. Thus, the rates of organotin monomer homopolymerization have been found to change in the order vinyltriphenylstannane > *p*-vinylphenyltricyclohexylstannane > vinylbisphenyltriphenylstannane.[44] At the same time, the activation energy of the process changes in an opposite manner. As the phenyl group is replaced by the cyclohexyl one, E_a increases; this is attributed to a diminished double bond side-chain conjugation in the case of vinylphenyltricyclohexylstannane. The extension of the side chain in the case of *p*-vinylphenylbisphenyltriphenylstannane lowers the reactivity under radical polymerization conditions.

New classes of organometallic polymers with a Sn—C bond in the side ester group have been synthesized by polymerization of 1-(trimethyltin)alkyl methacrylates.[15] Their polymerization is distinguished by chain transfer to an organotin polymer and a polymer due to the presence of a labile atom H in the α-position toward the $Sn(CH_3)_3$ group. Such a reaction is accompanied by a decrease in the polymerization rate, since the appearing polymer is incapable of participating in the initiation process. Note that the presence of a metal atom in the monomer molecule for the degenerate chain transfer is a must. As this takes place, the chain transfer process goes more readily to the polymer than to the MCM, since the growing radical abstracts the hydrogen atoms in a five-membered transition state:

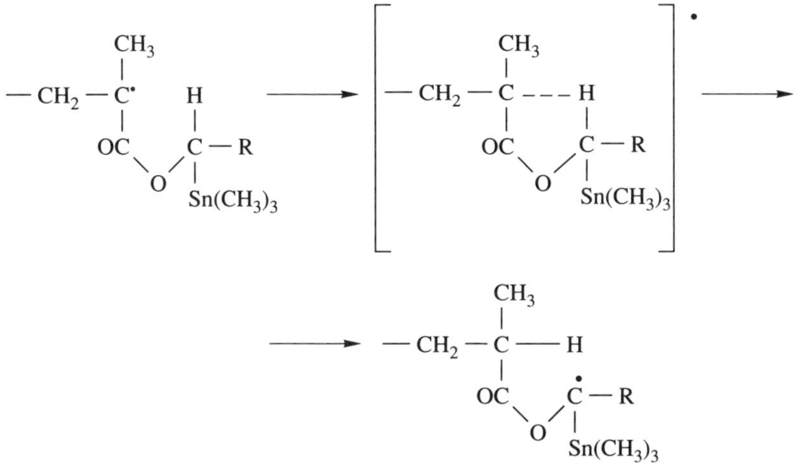

Such an intermolecular process is facilitated as compared with an intermolecular reaction of hydrogen atom abstraction as the chain is transferred to the MCM.

A moderate conversion is typical of solid-phase homopolymerization of a Hg(II) acetylene derivative (ultraviolet [UV] initiation at room temperature).[24] This is likely due to the fact that the appearing polymer is a kind of filter for the initiating radiation:

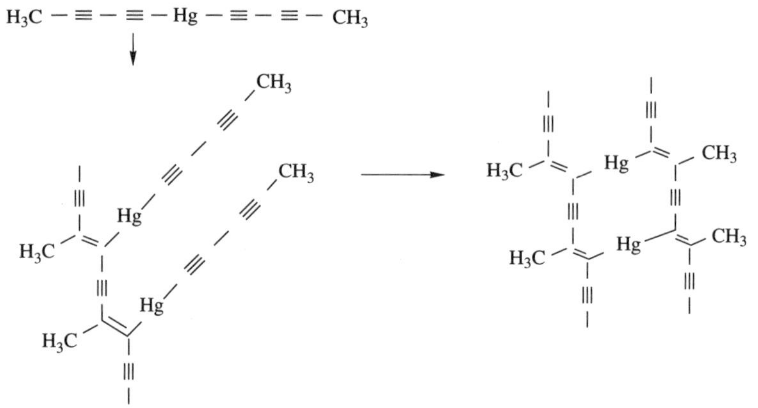

Finally, let us say a few words about the homopolymerization of organometallic styrene Pd(II) and Pt(II) derivatives:[25]

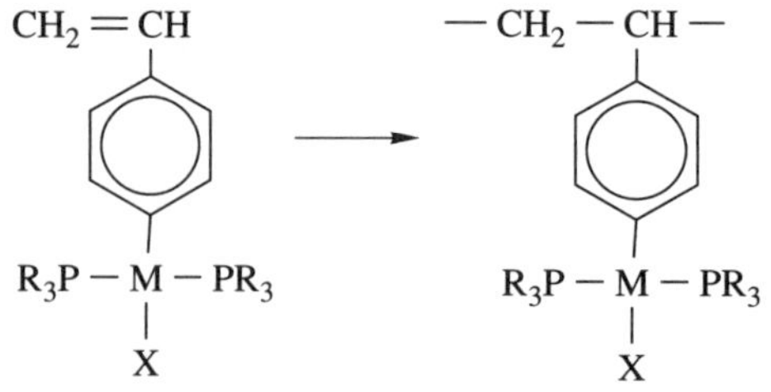

Pd(II)-containing monomers (X = Cl or Br) and also a Pt(II)-containing monomer are readily polymerizable in the presence of azobisisobutyronitrile (AIBN). At the same time, Pd(II)-containing monomers (X = CN and C_6H_5) decompose at a polymerization temperature of 55 to 60°C; however, they are distinguished by effective polymerization at 0 to 15°C when the process is initiated with $B(C_4H_9) \cdot O_2$. The MM of the resulting polymers is not high (P_n = 5.5 to 16.8). Attempts to conduct a cationic polymerization of these MCMs (BF_3 as the initiator) were not successful despite a favorably negative large value of the polar factor e equal to –1.4. The inhibition of cationic polymerization is attributed to the interaction of the palladium MCM fraction with BF_3.

B. HOMOPOLYMERIZATION OF COVALENT-TYPE MCMs WITH AN M—O BOND

Let us note first that a monomethyl methacrylic Mg(II) derivative of ethylene glycolate homopolymerizes in the presence of AIBN at 70°C, the yield of the product being as high as 35%.[53]

As can be seen in Figure 1, the polymerization activity of Ti(IV)-based monomers is appreciably dependent on the nature of the multiple bond-containing radical. The initiation reaction and the kinetic principles of the process have been studied in much detail with a particular reference to dimethylvinylethynylcarbinoloxytributoxytitanium.[45,46] The general equation for the polymerization of this MCM, $w_p = K[M]^{0.5}[I]^{0.5}$, is rather unusual for the radical polymerization of vinyl monomers. It can be assumed that such a depen-

Figure 1 Kinetic curves of homopolymerization yield of Ti(IV)-containing monomers $(C_4H9O)_3Ti(OR)$ with the following R:

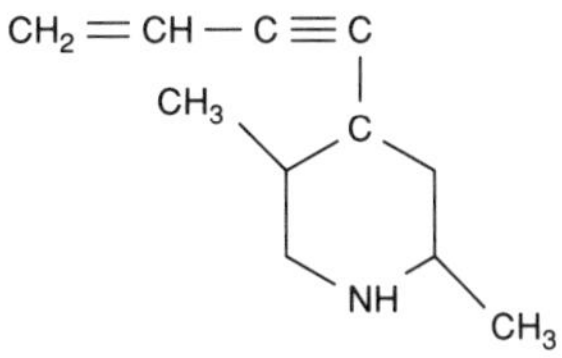

(1)

— $C(CH_3)_2C \equiv C - CH = CH_2$ (2)

— $CH_2CH_2OCOC(CH_3) = CH_2$ (3)

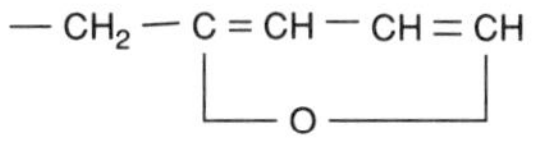

(4)

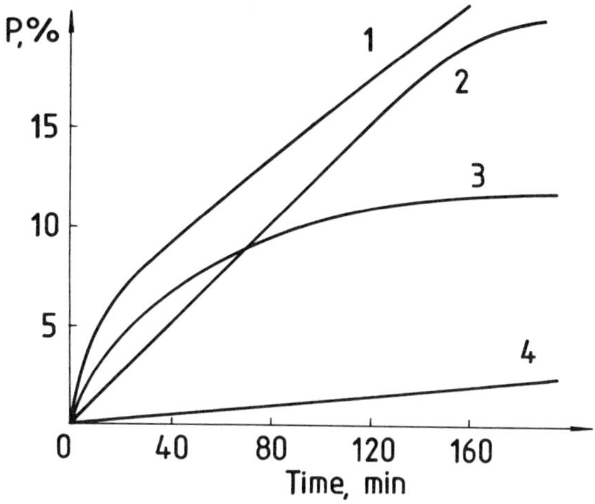

[MCM] = 2.8 mol/l, [AIBN] = $7.6 \cdot 10^{-2}$ mol/l; benzene, 80°C.

dence is one of the features characterizing the polymerization of these MCMs, particularly with respect to the interaction of primary radicals with a metal:

Chain initiation
$$I \xrightarrow{k_i} 2R^{\bullet}$$
$$R^{\bullet} + M \xrightarrow{k_k} RM^{\bullet}$$

Chain propagation
$$RM^{\bullet} + M \xrightarrow{k_p'} RM_n^{\bullet}$$
$$R^{\bullet} + M \xrightarrow{k_p} RM_n^{\bullet}$$

Chain termination
$$RM^{\bullet} + RM^{\bullet} \xrightarrow{k_t} \text{polymer}$$
$$RM^{\bullet} + R^{\bullet} \xrightarrow{k_t'} \text{polymer}$$
$$R^{\bullet} + R^{\bullet} \xrightarrow{k_t''} \text{polymer}$$

where k_k is the equilibrium constant of MCM-radical complexing; k_p' and k_p are the rates of chain growth reactions with and without Ti(IV)-coordinated and -uncoordinated radicals, respectively; k_t, k_t', and k_t'' are rate constants of the corresponding chain termination reactions.

The above scheme includes an MCM-radical interaction, whereas chain growth and termination can incorporate both the conventional and monomer-coordinated radicals. According to this scheme, the polymerization reaction equation will have the form:

$$w = k_p[RM][M] + k_p[R][M].$$

Based on the quasi-steady-state principle, the equilibrium for the current polymerization rate can be written as follows:

$$w = \left(k_p k_p'[M] + K'\right)\left(\frac{k_i I}{k_t' k_p'[M] + k_t''}\right)^{1/2} [M].$$

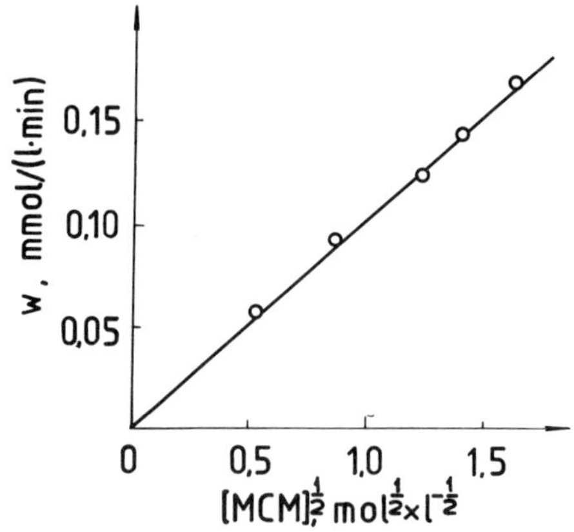

Figure 2 Graphic solution of equation for the dependence of initial rate of $(C_4H_9O)_3TiOC(CH_3)_2C{\equiv}C{-}CH{=}CH_2$ polymerization on MCM concentration. Benzene, 80°C, AIBN initiation.

For the initial polymerization rate at $k_p'' > k_p k_p'\,[M]$ and $k_t'\,k_p'[M] >> k_t''$, this equation has the following form:

$$w_o = k_p'\left(\frac{k_t I_0}{k_t' k_p'}\right)^{1/2}[M]_0^{1/2}$$

It is evident from Figure 2 that this equation agrees well with the experiment. Moreover, the solution of this equation at $t \to t_\infty$ and $[M] \to [M]_\infty$ describes well both the kinetics of polymer buildup:

$$[M]_0^{1/2} - [M]_\infty^{1/2} = k_p' \frac{I_0^{1/2}}{\left(k_t' k_p' k_i\right)^{1/2}}$$

and the dependence of the limiting conversion on the starting initiator concentration. The k_i value found from the last equation was equal to $1.3 \cdot 10^{-2}$ min; and the ratio of constants $k_p'/k_t'k_p'k_i$ as determined by various methods, is 1.10 to 1.28.

Special studies indicate that $Ti(OC_4H_9)_4$ does not affect the initiator decomposition rate. Consequently, the unusual "order" dependences in the course of radical polymerization of Ti(IV)-based MCMs are likely due to the following processes:

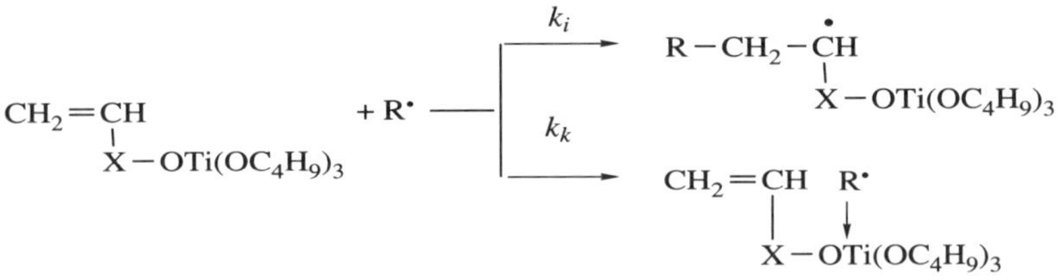

Along with the features resulting from chain propagation during polymerization of the MCMs under study, there may be some specific details of chain termination, such as an intramolecular one when the metal ion acts as a kind of a radical trap:

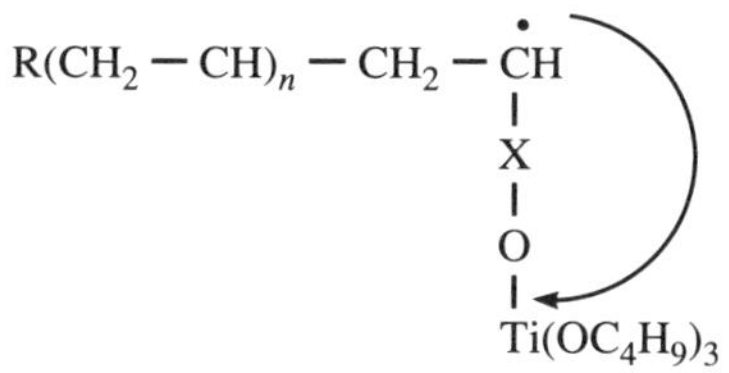

This entails relatively small MMs of the resultant polymers: at 25°C is equal to 0.1 to 0.05 dl/g.

Among the other Ti(IV)-containing monomers with a covalent M—O bond, mention should be made of the allyloxytitanocene chloride $(C_5H_5)_2Ti(OCH_2CH{=}CH_2)Cl$, the heating of which in the presence of 0.1% benzoyl peroxide (BP) gives rise to a trimer with MM = 770.[31]

Polymerization conversions are also typical of V(V)-containing monomers. Thus, monomethyl methacrylate ethylene glycolate derivatives, $OV(OR)_{3-n}(O\text{-}iso C_3H_7)_n$ of all the three compositions (AIBN initiation) are readily polymerizable with precipitation of a yellow polymer from the benzene solution.[47] The polymer yields are 6% at 65°C and as high as 40% at 75°C (MM = 13,000).

III. COPOLYMERIZATION OF COVALENT-TYPE MCMs

The situation in the field of covalent-type MCM polymerization is similar to that with homopolymerization. Despite the existence of a large number of published works, the quantitative data are scarce and qualitative findings predominate.

A. TRUE ORGANOMETALLIC MONOMERS AS COMPONENTS OF COPOLYMERIZABLE SYSTEMS

First of all, we shall describe the synthesis of copolymers containing a covalent-bonded Mg(II). Thus in the polymerization of styrene and α-methylstyrene in toluene, THF, and an HMPA-toluene mixture in the presence of *p*- and *m*-vinylbenzylmagnesium chloride, a small amount of the unreacted MCM is inserted in the polystyrene chain.[11] Styrene polymerizes on such side benzyl magnesium chloride groups to give rise to a branched metal-containing copolymer.

Vinyl and allyl Sn(IV) and Pb(IV) derivatives incapable, as mentioned above, of homopolymerization in ordinary conditions also produce an inhibiting effect on the radical polymerization of other monomers such as styrene, methyl methacrylate (MMA), and vinyl acetate (VA).[19,37,48] This can be attributed to the fact that the free radicals appearing in the system are trapped and bonded to the metal. As this takes place, the inhibiting effect increases with the number of unsaturated groups in the MCM molecule, the allyl compounds being more efficient inhibitors as compared with the vinyl ones. In terms of the inhibiting action, unsaturated Sn(IV) compounds are arranged as follows: $Sn(CH_2CH{=}CH_2)_4 > Sn(CH{=}CH_2)_4 > (CH_3)_3Sn(CH_2CH{=}CH_2) > (C_6H_5)_3Sn(CH_2CH{=}CH_2) > (C_6H_5)_3Sn(CH{=}CH_2)$. Copolymerization of such MCMs is distinguished by close to zero r_2 values (Table 3). The process occurs with low yields, and the final product contains normally a relatively small amount of MCM. Thus, the copolymerization of diallyl derivative of Ge(IV) and Sn(IV) with styrene and MMA is characterized by high yields not in excess of 69 and 42%, respectively; whereas the MCM content in the products does not exceed 18 mol%.[44] Triethylvinyl-lead-styrene and α-methylstyrene copolymers were found to contain 4.5 to 6 wt% of lead.[50]

As in the case of triethylvinylstannane homopolymerization, the copolymerization of this MCM with styrene and trimethylvinylsilane occurs smoothly when use is made of an

Table 3 **Relative reactivity constants and parameters *Q-e* of copolymerizable systems based on Sn(IV)- and Pb(IV)-containing covalent-type MCMs**

MCM (M_2)	Parameter	Comonomer: Styrene	Comonomer: MMA	Ref.
$(CH_3)_3SnCH{=}CH_2$	r_1	44.8	25.1	48
	r_2	0.001	0.03	
	Q_2	0.005	0.036	
	e_2	0.96	0.93	
$(C_4H_9)_3SnCH{=}CH_2$	r_1	16.0	27.9	48
	r_2	0.005	0.03	
	Q_2	0.017	0.03	
	e_2	0.82	0.82	
$(CH_3)_3SnCH_2OCOC(CH_3){=}CH_2$	r_1	–	1.0	15
	r_2	–	1.0	
	r_1	0.826	–	49
	r_2	2.86	–	
	r_1	0.96	–	49
	r_2	1.6	–	
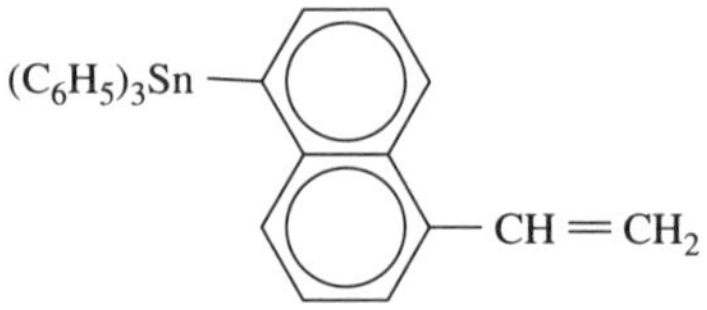	r_1	0.897	–	49
	r_2	4.7	–	
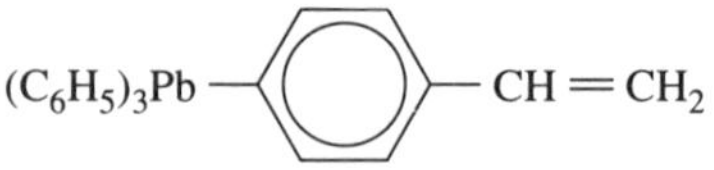	r_1	0.978	–	49
	r_2	1.22	–	

anionic initiator, namely, C_4H_9Li.[40] The MM values of these copolymers are 6000 to 9000.

Styrene Sn(IV) and Pb(IV) derivatives are much more active in these processes.[49] As shown in Table 3, the activity of the Pb(IV)-containing monomer is close to that of the styrene- but lower than that of the Sn(IV)-based monomer. In terms of activity in copolymerization with styrene, the organotin monomers form the series $(C_6H_5)_3SnC_6H_4CH{=}CH_2 > (C_6H_5)_3SnC_{12}H_8CH{=}CH_2 > (C_6H_{11})_3SnC_6H_4CH{=}CH_2$. It is evident that the increase in the size of a substituent and a weaker double bond conjugation lead to a lower activity of these MCMs. Note also the possibility of producing synthetic organotin rubbers as a result of styrenetrimethylstannane-butadiene copolymerization.[19]

Some interesting features are typical of the copolymerization of another organotin monomer, 1-(trimethyl)methyl methacrylate, with MMA.[15] The process is distinguished by the fact that the composition of the copolymer is close to that of the parent monomeric mixture (r_1 = r_2 = 1.0). In other words, in this case one can observe an "azeotropic" copolymerization. Thus, the substitution of an organometallic group for the hydrogen atom does not affect the double bond reactivity, which is likely due to the presence of a three-atom bridge.

A more detailed consideration has been given to the copolymerization of styrene with its Pd-containing derivative, *trans*-$Pd[P(C_4H_9)_3]_2(C_6H_4CH{=}CH_2)Cl$.[25] The shape of the copolymer composition curve and the values of the relative reactivity constants ($r_1 = 1.49$ and $r_2 = 0.45$) indicate that this MCM is a less active monomer in copolymerization as compared with styrene. Note a large negative e_z value equal to –1.4. It can be concluded that the σ-bonded *p*-$Pd[P(C_4H_9)_3]_2$ group possesses strong electrodonor properties, the electronic density in the MCM being heavily shifted to the C=C bond.

B. COPOLYMERIZATION OF COVALENT-TYPE MCMs WITH AN M—O BOND

Data on the copolymerization of alcohol-derived MCMs are extremely scant and mainly empirical. An exception to this is a study of diisobutylaluminumalkoxy monomer-butadiene copolymerization in hydrocarbon environments on a neodimium catalyst.[26] It is a very rare case of the successful application of the Ziegler-Natta catalyst to polymerization MCM conversions. Copolymerization gives rise to a polar group-modified *cis*-polybutadiene. Aluminum content in the copolymer increases with MCM fraction in the same copolymer (Figure 3). As this takes place, the fraction of 1,4-*cis*- in the microstructure of the butadiene part of the copolymer decreases from 94.4 to 88.9% as a result of increasing the 1,4-*trans*-fraction (from 5.1 to 10.5%) at the same content of the 1,2 units (0.5 to 0.6%). The MCM is believed to be distributed along the chain in the form of isolated units, each being preceded by a *trans*-structured butadiene one.

As with Ti(IV)-containing covalent-type MCMs, tetraallyloxy titanate is incapable of homopolymerization.[51] However, in the absence of any initiators at room temperature this MCM can spontaneously copolymerize with VA to yield a nonmeltable substance insoluble in standard organic solvents. Note that the copolymer of this MCM with MMA was obtained in the presence of BP at 65°C. Other authors report on an effective copolymerization of allyloxytitanocene with styrene and MMA (in the presence of BP as the initiator, 100 to 120°C).[52] The MM of the styrene-containing copolymer is 20, 100 against 50,000 of polystyrene obtained in the same conditions. A similar decrease in the copolymer MM is observed in the copolymerization of this MCM with MMA (70,000 and 90,000, respectively).

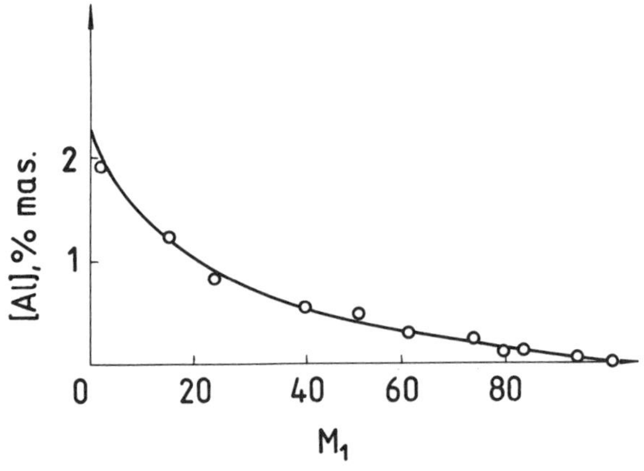

Figure 3 Content of Al(III) in butadiene and $CH_2{=}C(CH_2{-}CH_2OAl(\textit{iso}\text{-}C_4H_9)_2){-}CH{=}CH_2$ copolymer (M_2) as a function of monomer mixture composition.

Note, finally, that in the course of devitrification of the irradiated matrices in THF it has been possible to copolymerize two different covalent-type MCMs.[53]

$$CH_2{=}CH{-}C{\equiv}C{-}C(CH_3)_2OTi(OR)_3$$

and

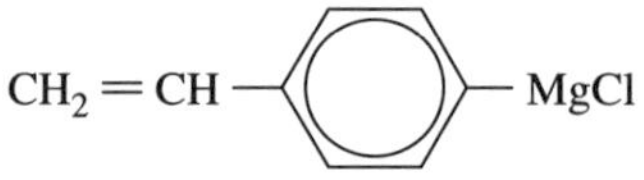

The copolymer yield is 55%, and the product contains 1.8 wt.% of Ti and 5.3 wt.% of Mg.

Concerning the V(V)-containing monomers, there are reports on the formation of copolymers based on benzene-soluble monomethyl methacrylic ethylene glycolate with styrene.[47]

IV. GRAFT POLYMERIZATION OF COVALENT-TYPE MCMs

Graft polymerization can be regarded as a conventional radical (less frequently ionic) polymerization, except that the initiating particles are fixed on the substrate of the polymer or any other material. This process can be interpreted as a special type of block polymerization.

Unfortunately, there are no published data on the possibility of implementing graft polymerization of true organometallic monomers. At the same time, some covalent-type MCMs with an M—O bond can be effectively involved in a graft reaction onto powdered polyethylene (PE) and polypropylene (PP).[35] Given below is the scheme of this process:

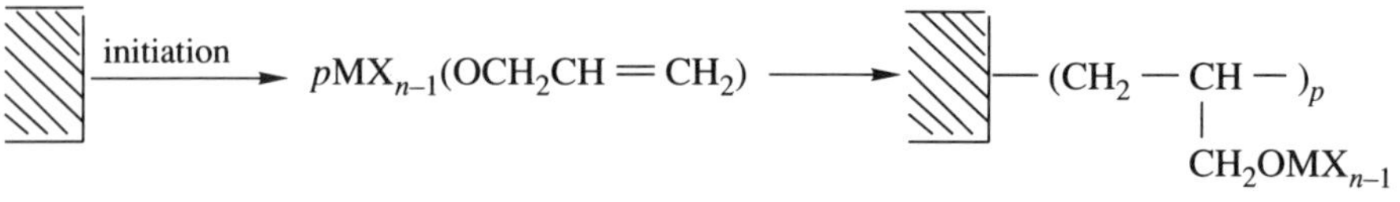

To initiate the reaction, the polymeric substrate was preirradiated *in vacuo* using ^{60}Co, accelerated electrons, and a high-frequency discharge. The irradiated polymer was brought in contact with an MCM solution in benzene, toluene, and heptane at 75 to 80°C. Used as the MCMs were allyl alcohol derivatives such as $Ti(OCH_2CH{=}CH_2)\ (OC_3H_7)_3$, $Cl_3Ti(OCH_2CH{=}CH_2)$, $(C_5H_5)_2Ti(OCH_2CH{=}CH_2)Cl$, $(C_4H_9)_2Ti[NH(CH_2CH{=}CH_2)]_2$ and $(C_2H_5O)_2VO(OCH_2CH{=}CH_2)$. Metal content in the graft polymers does not exceed 0.12 wt.% which is not suprising considering a small tendency of allyl monomers toward polymerization conversions.

Despite the fact that the initial information on covalent-type MCM polymerization conversions appeared 50 years ago, generally it can be concluded that the experimental data available are very scattered. Exceptions to this are the results of studies that date back to the 1950s and the early 1960s. They were concerned with the polymerization properties of vinyl, allyl, and styrene derivatives of Ge(IV), Sn(IV), and Pb(IV). However, later on these works were discontinued; and in the last few decades only isolated attempts were seen in this field. In this context, researchers are encouraged by the appearance of new studies on the synthesis and polymerization of covalent-type MCMs (cf., for example, Reference 11). Much new work is to be done in the future.

Chapter 2

Ionic-Type Metal-Containing Monomers

This class of monomers comprises, largely, salts of unsaturated mono- and dicarboxylic acids and has found extensive applications. The simplicity of the experimental setup, mild synthesis conditions, availability of starting reagents, and comparative ease of isolating the products and their stability make these (metal-containing monomers) MCMs among the most convenient species.

Mention should be made of the multiple functions of the $RCOO^-$ carboxylate group.[54] Upon coordination with the metal ion it can act as a mono-, bi-, tri-, and even tetradentate ligand. However, the most frequently encountered case is when $RCOO^-$ represents a mono- (*a*) and a bidentate (*b*) ligand:

The coordination type can be proved by both X-ray structure and infrared (IR)

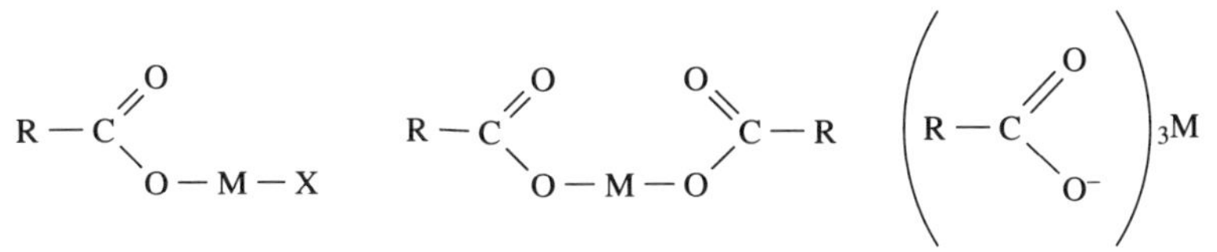

(*a*)

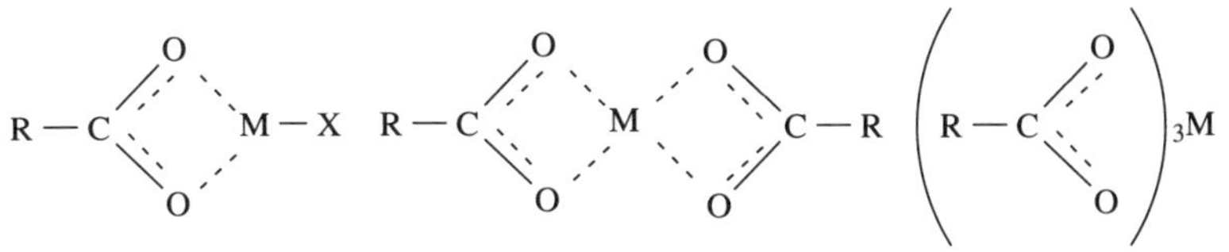

(*b*)

spectroscopy: for structures (*b*) ν_{COO^-} appears at 1560 to 1580/cm.[55]. Bands beyond 1600/cm and at 1700/cm usually indicate the absorption of a carbonyl group within a monodentate-coordinated and an undissociated carboxylate group, respectively.

I. SYNTHESIS OF IONIC-TYPE MCMs

A. ACRYLATES AND METHACRYLATES OF METALS BELONGING TO THE MAJOR GROUPS

The syntheses of these MCMs have been studied very well. Alkali metal methacrylates can be readily obtained through interaction of acrylic acid (AA) and methacrylic acid (MAA) with the hydroxide of the corresponding metal in methyl and ethyl alcohol.[56–58] Equally applicable are metal carbonates, e.g., in the case of Rb acrylate synthesis. The resulting MCMs are precipitated with ether or acetone. Alkali metal methacrylates are anhydrous salts, although they can give crystalline adducts with excess acid reagent, such as $CH_2{=}C(CH_3)COOH \cdot CH_2{=}C(CH_3)COOK$.[59]

Methacrylates of rare-earth metals are produced similarly, by interacting $MgO[Mg(OH)_2{\cdot}MgCO_3{\cdot}3H_2O]$, $CaO(CaCO)_3$, $Ca(OH)_2$, $Sr(OH)_2$ and $Ba(OH)_2 \cdot 8H_2O$ with AA or MAA in water or methanol.[60–64] The product thus obtained is precipitated with excess methanol, ether, or acetone, or—in the case of a reaction taking place in water—

by crystallization from solution with slow water evaporation. These MCMs comprise crystalline hydrates with different amounts of coordinated water. Thus, Sr(II) acrylate is a semihydrate; Mg(II) and Ca(II) acrylates are mono- and dihydrates, respectively; and Ba(II) acrylate and methacrylate give rise to mono-, di- and even pentahydrates.[61–63,65–68]

The dehydration of such MCMs has been much discussed due to the great effect produced by crystal water on their polymerization properties. An anhydrous Mg(II) acrylate can be obtained from the monohydrate being heated at 100 to 150°C.[65] A Ca(II) acrylate dihydrate is dehydrated by heating *in vacuo* at 60°C,[61, 67] as well as in air at 100 to 150°C.[65, 69] In the latter case the process occurs in two stages, by abstracting first 1.5 mol and then 0.5 mol of water at temperatures under and over 100°C, respectively. Also, two stages are required to dehydrate a Ba(II) methacrylate monohydrate: (1) transition to the other crystal form (anhydrous salt) and (2) dehydration proper at 65 to 140°C; its decomposition starts at 210°C.[70] An anhydrous Ba(II) methacrylate can be produced by heat-treating a crystalline hydrate at 47 to 50°C.[62,68,71] One more way for dehydrating rare-earth metal methacrylates is through recrystallization from anhydrous methanol[69,71] as well as by synthesis at 40 to 100°C in hydrocarbon solvents with azeotropic removal of water and subsequent drying.[64]

An anhydrous Ba(II) methacrylate is one of the few MCMs that has been studied by an X-ray method.[72] Its crystals are monoclinic (a = 12.162, b = 8.017, c = 10.373 Å); the CN of Ba ions is equal to 8; and a coordination polyhedron is the irregular type having oxygen atoms at the vertices with the Ba—O distances equal to 2.766 to 3.030 Å and localized double bonds (1.33 and 1.34 Å). The barium atoms are positioned near the X = 0 plane and are interconnected via methacrylate ligands.

Information of Al(III) methacrylates can be mainly found in patent publications. The methods of their synthesis are similar to those described above. Thus, an $Al(OH)_3$ gel is caused to interact with a mixture of saturated and unsaturated carboxylates. At the same time, organoaluminum compounds react with AA and MAA due to the presence of the mobile H of the carboxyl groups, which is relied on in the production of the corresponding MCMs.[73–76]

$$CH_2{=}C(R)COOH + Al(C_2H_5)_2Cl \rightarrow CH_2{=}C(R)OCOAl(C_2H_5)Cl + C_2H_6$$
$$CH_2{=}C(R)COOH + Al(C_2H_5)Cl_2 \rightarrow CH_2{=}C(R)OCOAl(C_2H_5)Cl + HCl$$

A similar method can be applied to the production of the corresponding derivatives of higher unsaturated acids.[76]

Note also the synthesis of methacrylic Hg(II) and Ge(IV) derivatives $C_6H_5HgOCOC(R){=}CH_2$[77] and $(C_2H_5)_3GeOCOC(CH_3){=}CH_2$[78] (R = H or CH_3).

Tin- and lead-containing methacrylates have enjoyed wide use in polymerization practices.[9] Such MCMs can be divided into two types: salts of unsaturated acids, $M(OCOC(R){=}CH_2)_2$, (R = H or CH_3); and their organic tin (lead) esters. Then Pb(IV) methacrylates are obtained by reacting lead salts, oxide, or carbonate with AA or MAA in water, as well as in benzene, tetrahydrofuran (THF), etc. The same method is used to obtain Sn(IV) methacrylates. However, a heavier reliance is put on other types of MCMs, namely, organometallic AA and MAA esters:

$$R_3'M(OCOC(R){=}CH_2) \text{ and } R_2'M(OCOC(R){=}CH_2)_2$$

(R = alkyl or aryl).

Such MCMs are synthesized by interacting trialkylstannanols R_3SnOH, triphenyl-plumbanol $(C_6H_5)_3PbOH$, and polyalkyl(aryl)stann(plumb)oxanes $(R_2MO)_x$, where M = Sn(IV), Pb(IV), as well as hexalkyldistannoxanes $R_3SnOSnR_3$ with AA or MAA:

$$R_3SnOH + CH_2{=}C(R)COOH \rightarrow R_3Sn(OCOC(R){=}CH_2) + H_2O$$
$$R_3SnOSnR_3 + 2CH_2{=}C(R)COOH \rightarrow 2R_3Sn(OCOC(R){=}CH_2) + H_2O$$
$$(C_6H_5)_3PbOH + 2CH_2{=}C(R)COOH \rightarrow (C_6H_5)_2Pb(OCOC(R){=}CH_2)_2 + C_6H_5 + H_2O$$

To enhance the yield of the product, AA and MAA can be replaced by the corresponding anhydride.[79] Interestingly, parts of such MCMs (higher di- and tri-alkyltin methacrylates) are viscous liquids and other parts are waxy substances.

The structure of organotin methacrylates is characterized by association of the carboxylate groups contained therein:[80]

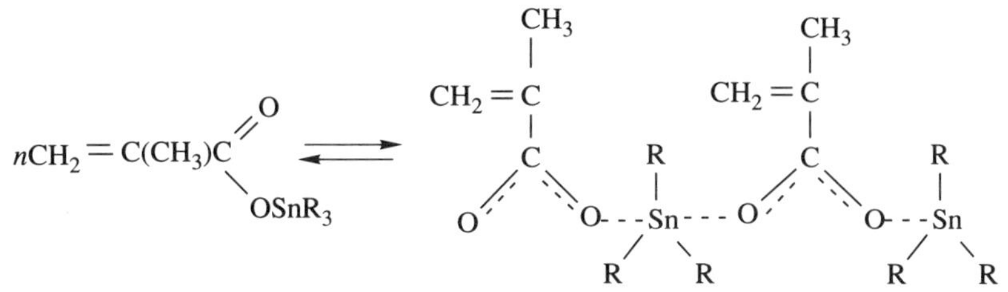

where R = CH_3, C_3H_7, or C_4H_9.

As the length of the alkyl substituent at the tin atom increases, the ν_{c-c} intensity grows, on the one hand; and ν_{c-o} in the carboxylate ion shifts to the high-frequency region 1590 to 1620/cm, on the other hand, to suggest the effect of the SnR_3 group on the vibrations of the C—O group conjugated with the multiple bond.

Note also the synthesis of a Sn(IV)-containing ionic-type monomer with the acrylate group being moved away from the metal ion at a considerable distance:[81]

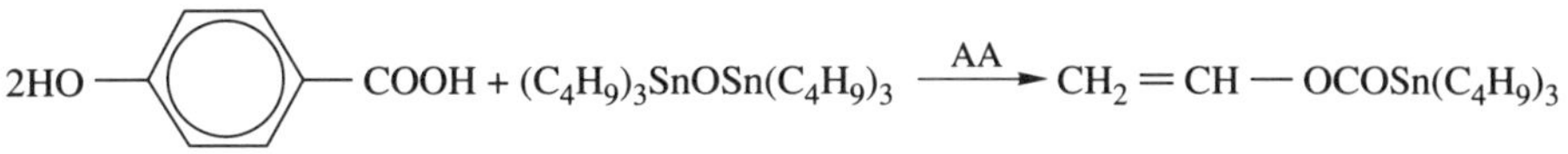

B. TRANSITION METAL (METH)ACRYLATES

The interaction of hydroxides or hydrocarbonates with unsaturated acids is a simple and convenient method of synthesizing methacrylates of most metals in the first transition series. A general approach underlying this method is the effect of excess acid both in the absence of solvent[82] and in methanol and ethanol solutions, as well as in hydrocarbon suspensions.[83] The latter approach has been used to advantage for producing Co(II), Ni(II), Zn(II), and Cu(II) acrylates. The process is distinguished by a high yield (over 95%) of the product containing more than 94% of double bonds with respect to the calculated value. The resulting Cu(II) and Zn(II) acrylates comprise anhydrous salts whereas the Co(II) and Ni(II) acrylates contain crystal water. The absence of crystal water in the Zn(II) acrylate molecule also has been demonstrated by special studies.[65] Electronic spectral studies of Co(II) and Ni(II) acrylates have been useful in revealing their octahedral coordination,[84] which is consistent with the effective magnetic moment value (Table 4). These MCMs have the following structure:

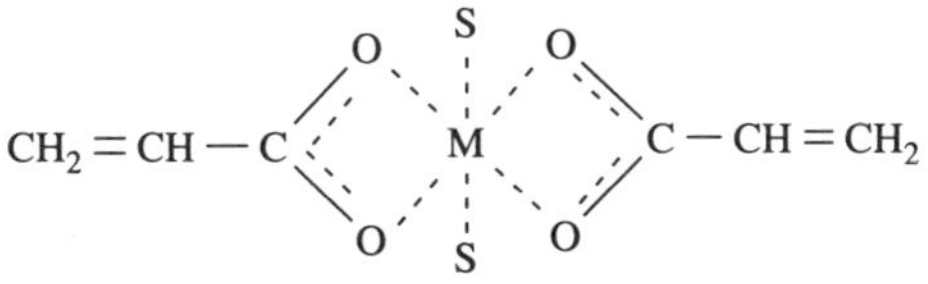

Table 4 **Some characteristics of transition metal acrylates**[83,84]

MCM	Characteristic IR spectrum frequencies (per cm^{-1}) $\nu_{C=C}$	ν_{COO^-}	$\nu_{M—O}$	Magnetic moment (μ_{ef}, μ_B) 21°C	–195°C
$(CH_2{=}CHOCO)_2Co{\cdot}H_2O$	1640	1560	280	5, 10	4, 53
$(CH_2{=}CHOCO)_2Ni{\cdot}H_2O$	1640	1560	290	3, 60	3, 47
$(CH_2{=}CHOCO)_2Cu$	1645	1575	315	1, 40	0, 22
$(CH_2{=}CHOCO)_3Fe$	1640	1520	400	–	–
$(CH_2{=}CHOCO)_3Cr$	1639	1535	440	–	–
		1580			
$(CH_2{=}CHOCO)_2Mn$		1585		–	–
$(CH_2{=}CHOCO)_4Mo$	1640	1480		–	–

The ionicity of the M—O bond in Co(II) and Ni(II) acrylates is 0.84 to 0.88. This points to a decrease of the positive charge at the ion metal due to the inductive ligand effect and suggests a marked orbital contribution to the metal-ligand bond.

In polar solvents such as methanol, ethanol and dimethyl formamide (DMFA) transition metal acrylates are weak electrolytes even in very dilute solutions (10^{-5} to 10^{-4} mol/l), their dissociation constant being $(1.57 \text{ to } 2.35){\cdot}10^{-5}$.[85]

Standard methods have been applied to the description of Mn(II) and Cr(III) acrylates.[86] There are reports on the production of Mn(II), Co(II), Ni(II), Cu(II), and Fe(III) methacrylates by interacting metal acetate with MAA,[87] although one can hardly suspect a quantitative yield of the product.

Note also the synthesis of Co(III) methacrylate derivatives, namely, methacrylatopenta- and *cis*-dimethacrylatotetraamminocobalt perchlorates:[88]

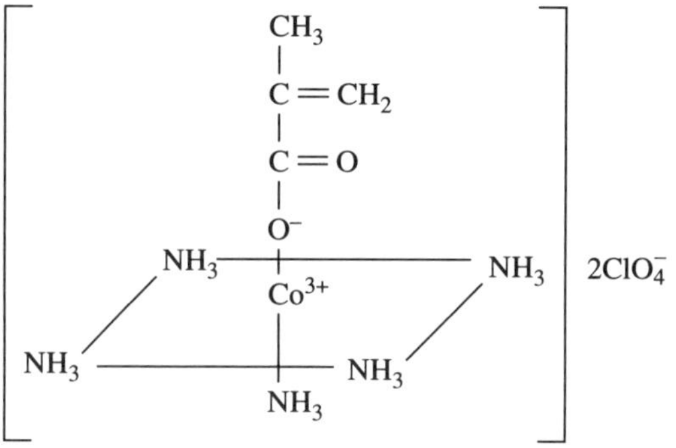

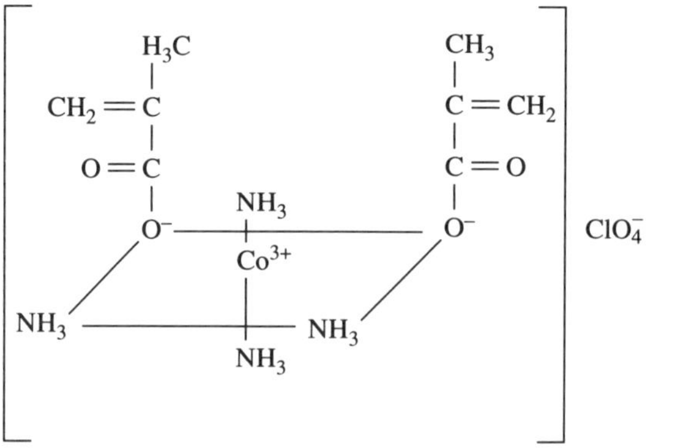

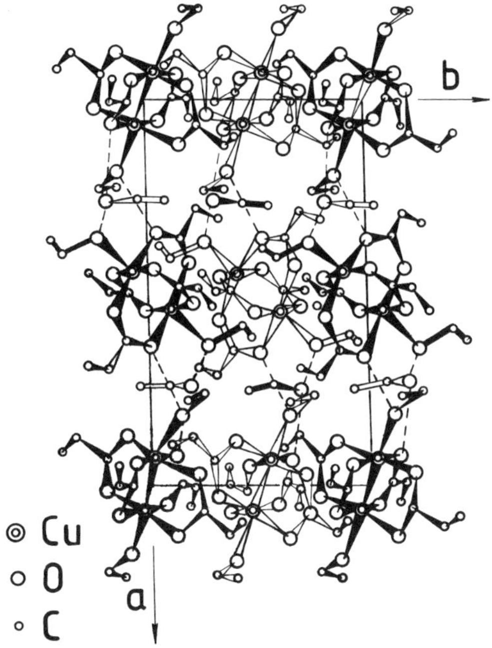

Figure 4 Crystalline structure of copper acrylate $[Cu_2(OCOCH{=}CH_2)_4(C_2H_5OH)_2] \cdot C_2H_5OH$.

Among the transition metal acrylates described above, the most detailed consideration has been given to a Cu(II)-containing MCM structure.[84] Its X-ray electronic spectrum contains—along with the base $Cu(2p_{3/2})$ line—an intensive satellite on the higher coupling energy side, the appearance of which is linked with the presence of a lone electron at the radiating site. Due to the antiferromagnetic exchange, its intensity is lower than that of copper oxide. Blue-green single crystals $[Cu_2(OCOCH{=}CH_2)_4(ROH)_2]ROH$, where $R = CH_3$ or C_2H_5, can be isolated from the solutions of this MCM in ethanol or methanol. For an ethanol-containing complex the following crystallographic evidence was obtained: $a = 15.745$, $b = 9.164$, $c = 17.826$ Å, $\gamma = 88.82°$, $V = 2572$ Å^3. A projection along the monoclinicity axis of this single crystal is shown in Figure 4. An interesting feature of its structure consists of the formation of a binuclear cluster with the Cu—Cu distance equal to 2.609 Å. In this case the copper atoms are interconnected via carboxylate bridges to form a complex complemented with two ethanol molecules. Studies of the nearest atomic surroundings of copper indicate that its coordination can best be described as a square-pyramidal one.

Of great interest are mixed salts of unsaturated and saturated acids, such as $CH_2{=}CHOCO{-}M{-}OCOCH_2CH_3$. Two ways have been tested in the synthesis of such MCMs. The first one is through neutralization of the corresponding hydroxides or hydrocarbonates with equimolar mixtures of the appropriate acids, although in this case the appearance of mixed products due to a differing strength of the reacting acids is likely. The second way involves selective hydration of one of the multiple bonds, e.g., in Co(II) and Ni(II) acrylates.[89]

Methods of synthesizing ionic-type Ti(IV)-containing MCMs closely resemble those used to obtain unsaturated alkoxy derivatives. First of all, this is a reetherification process. The products yielded by $Ti(OR)_4$ and AA or MAA are viscous, relatively fast hydrolyzable liquid or waxy substances readily soluble in standard organic solvents. Production of both mono- and polymethacrylic substituted products is possible. Usually, alkyloxy groups are substituted by a direct interaction of AA or MAA with $Ti(OR)_4$.[90] However, this may involve side processes such as etherification between the starting acid and the resultant alcohol and formation of oxide compounds of the type $TiO_x(OCOR)_{4-2x}$, $x = 0.5$ or 1.0.[27] Therefore, it has been found that insertion of unsaturated acyl residues into alkoxides using anhydrides of the corresponding acids is preferred:[91]

$$Ti(OR)_4 + n(CH_2=\underset{\underset{\displaystyle CH_3}{|}}{C}-CO)_2O \longrightarrow (CH_2=\underset{\underset{\displaystyle CH_3}{|}}{C}-COO)_n\,Ti(OR)_{4-n} +$$

$$nCH_2=\underset{\underset{\displaystyle CH_3}{|}}{C}-COOR$$

Isolation of these MCMs is quite a problem. Thus, equimolar amounts of $Ti(O\text{-}iso\text{-}C_3H_7)_4$ and methacrylic anhydrate in ether yield a crystal adduct:

$$(iso\text{-}C_3H_7O)_3TiOCO\underset{\underset{\displaystyle CH_3}{|}}{C}=CH_2 \cdot CH_2=\underset{\underset{\displaystyle CH_3}{|}}{C}-\underset{\underset{\displaystyle O}{\|}}{C}-O-iso\text{-}C_3H_7$$

This is attributed to a stronger partial charge at the titanium atom as the alkoxy groups in $Ti(OR)_4$ are substituted by more electronegative methacrylic ones, $(RO)_3Ti—O—CO—C(CH_3)=CH_2$.[92] A large difference in titanium and methacrylic group electronegativities is responsible for a marked ionicity of the Ti—O bond in such compounds (as high as 0.59 to 0.67).[93] As the carboxyl group is coordinated with a titanium atom, this effect can be much stronger so that the carboxyl group in such compounds will be ionized and exist in the form of a carboxylate anion.

Similar methods are equally applicable to the synthesis of ionic-type MCMs with titanocene dichloride:[94–96]

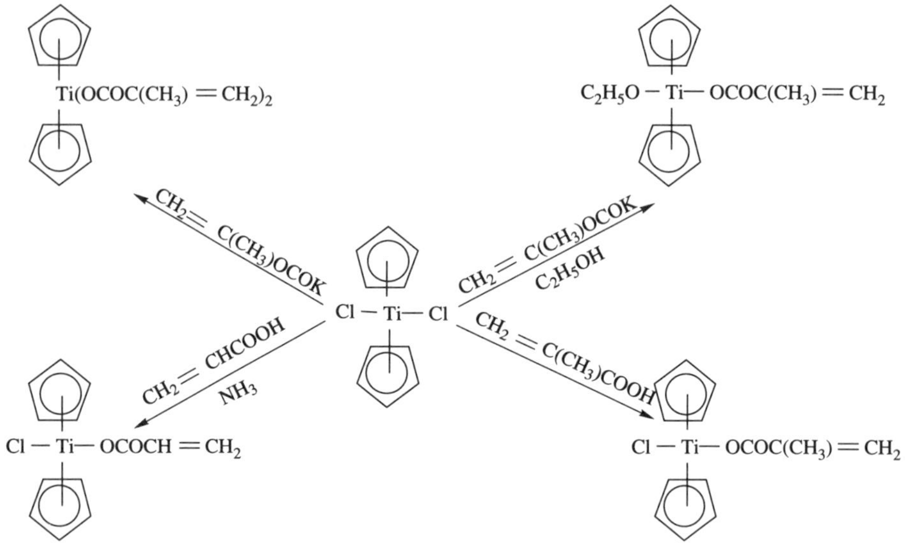

Mono- and dimethacrylic titanocene derivatives are well soluble in benzene, acetone, DMFA, pyridine, and even methyl methacrylate (MMA).[96]

Recently, much attention has been paid to the production of MCMs using rare-earth and transuranium elements. The reaction of AA or MAA with hydroxides gives La(III), Eu(III), Y(III), Pr(III), Gd(III), Tb(III), and Nd(III) methacrylates[97–99] in alcohol and water; and that of MAA with UCl_4 in toluene affords a methacrylic derivative of uranyl dichloride, $(CH_2{=}C(CH_3)OCO)_2UCl_2$.[100] Moreover, there are reports on the production of uranyl acrylate $UO_2(OCOCH{=}CH_2)_2$.[101] Derivatives of a higher unsaturated (undecylenic) acid in the form $M[OCO(CH_2)_8CH{=}CH_2]_3$, where M = Eu(III) or Tb(III), have also been synthesized.[102]

Note that methacrylates of most transition metals are very convenient objects of studies. They are well soluble in polar solvents, especially in anhydrous salts; and are distinguished by stability, particularly in air. Thus, minor changes of the double bond content in Co(II), Ni(II), Cu(II), and Zn(II) acrylates appear no sooner than in 8 to 10 months of storing in atmospheric conditions.[83]

C. METAL-CONTAINING MONOMERS BASED ON SUBSTITUTED UNSATURATED ACIDS

This is a very important class of MCMs that can be listed among "surfactant"-type monomers.[103] Such MCMs are distinguished by the presence of an ionogenic group (usually, a sulfonate one) and often contain hydrocarbon bridges. Usually alkali and, less frequently, alkali-earth metals are used. These monomers are largely represented by vinyl-, *n*-styrene-, and allyl sulfonates and other species with a more sophisticated structure such as:

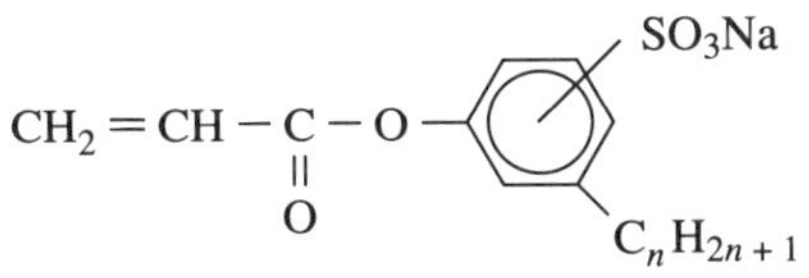

The methylene bridges spanning the double bond and the sulfonate group can be inserted as follows:[104, 105]

$$HO(CH_2)_6OH \xrightarrow{HCl} HO(CH_2)_6Cl \xrightarrow{Na_2SO_3}$$
$$\rightarrow HO(CH_2)_6SO_3Na \xrightarrow{CH_2=C(CH_3)COOH}$$
$$\rightarrow CH_2 = C(CH_3)OCO(CH_2)_6SO_3Na.$$

Among other MCMs based on substituted unsaturated acids mention should be made of Sn(IV), Pb(IV), Sb(III), and Hg(II) derivatives of vinylbenzoic acid:[106]

$$HOOCC_6H_4CH{=}CH_2 + (C_6H_5)_nMOH \rightarrow (C_6H_5)_nMOCOC_6H_4CH{=}CH_2 + H_2O$$

where n = 1 to 3.

Also, the MCMs under discussion should be complemented with such an interesting class of compounds as diacetylene derivatives $[CH_3\text{—}(CH_2)_{17}\text{—}C{\equiv}C\text{—}C{\equiv}C\text{—}COO]_nM$, where n = 1, 2; M = Na(I), Ag(I), Cd(I), Cu(II), Hg(II), and Mn(II)[107] obtained in single and multilayers from the appropriate acid in aqueous solutions.

Obviously, the list of possible MCMs based on substituted unsaturated acids can be enlarged.

D. METAL-CONTAINING MONOMERS BASED ON UNSATURATED DICARBOXYLIC ACIDS

In this group of MCMs the largest recognition has been gained by salts of maleic and fumaric acids. Depending on the type of metal and relationships between the reagents and the reaction conditions, maleic acid derivatives can be obtained in the form of chelates as well as monosubstituted or linear associates:

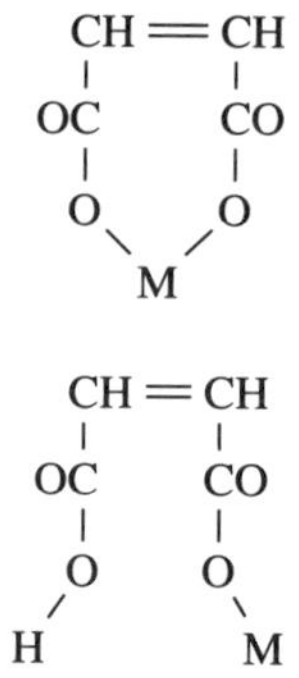

$$pHOOCCH{=}CHOCOMX_{n-1} \rightarrow$$
$$\rightarrow \ldots HOOCCH{=}CHOCOMX_{n-1} \ldots$$

Salts of alkali metals and maleic acid are formed by neutralizing its aqueous solution or anhydride with 10% excess alkali.[108] A single crystal of potassium fumarate dihydride $K_2H_2C_4O_4 \cdot 2H_2O$ has been identified.[109] In water environment maleic acid reacts with ZnO as a monobasic substance to yield a zinc salt which, due to intramolecular bonds and the presence of water, gives fibrous anizotropic products. Upon water removal, these products convert to powder.[110] X-ray structure studies indicate that in the crystal of this MCM, as well as of Co(II) and Ni(II) maleates, each metal atom is associated with one rather than two maleic acid residues. In other words, these are acid salts. The zinc atoms occupy positions at the inversion centers, and their coordination is complemented with four water molecules to form an octahedron:[111]

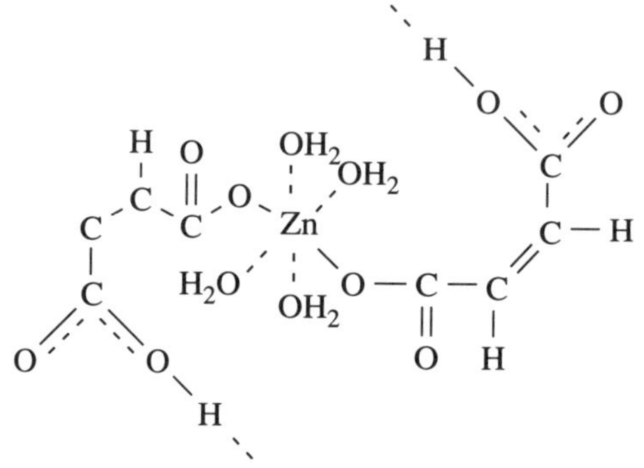

It is interesting that Zn(II) maleate can give rise to a polymerizable complex with ethylene diamine, $Zn[OCOCH{=}CHCONH(CH_2)_2NH_2]_2 \cdot 4H_2O$.[112]

X-ray structure studies of crystals of the cadmium maleate dihydride $Cd(C_4H_2O_4) \cdot 2H_2O$ suggest that their lattice is monoclinic (a = 6.08, b = 16.39, c = 7.00 Å, CN = 4).[113]

Stannols, plumbanols, and alkyl(aryl)stannoxanes (plumboxanes) interact readily with maleic anhydride to give the corresponding substituted maleates comprising solid substances crystallizable in the form of needles or plates soluble in organic solvents:[9]

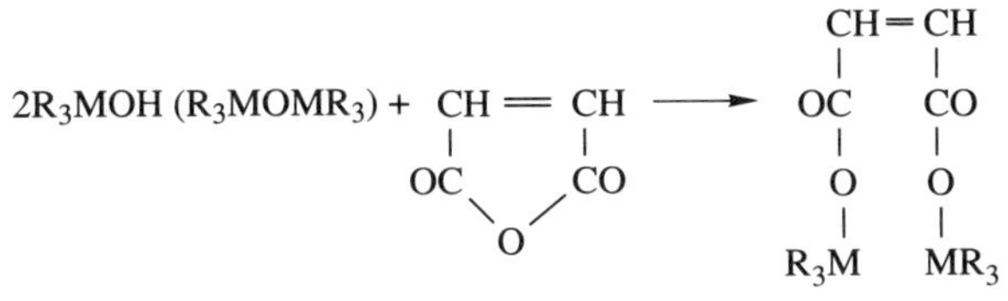

Also, a reaction of $(C_4H_9)_2SnO$ with maleic anhydride gives a chelate derivative:[114]

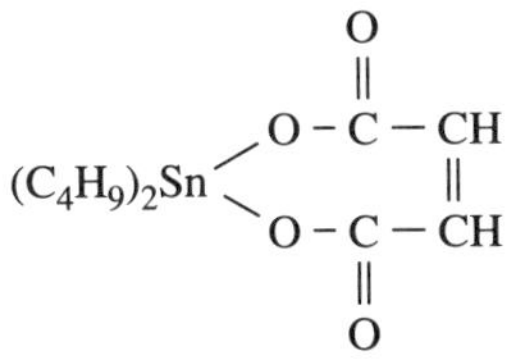

A waxy tributyltinacetate maleimide has been synthesized from maleic anhydride:[115]

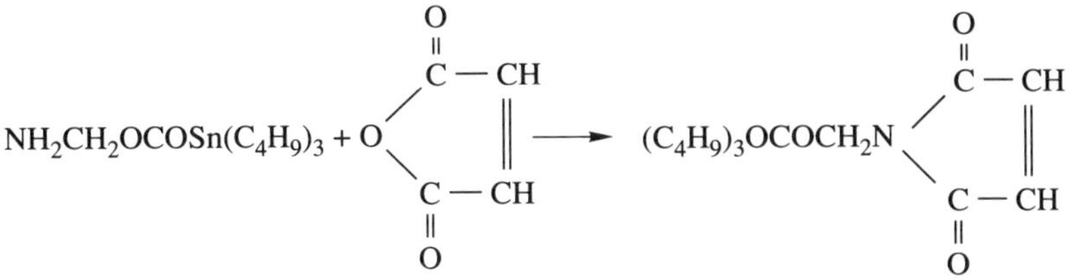

One more group of polymerizable Sn(IV)-containing derivatives of another dibasic (itaconic) acid has been made available:[116, 117]

Titanium(IV)-containing monomers under study are represented by acyl derivatives obtained from alkyl orthotitanates. Thus, oligomeric dibutoxytitanium monomaleate, a yellow-colored liquid, has been isolated.[90] The existence of intramolecular coordination with the appearance of chelate structures in such compounds has been confirmed by spectral studies:[93]

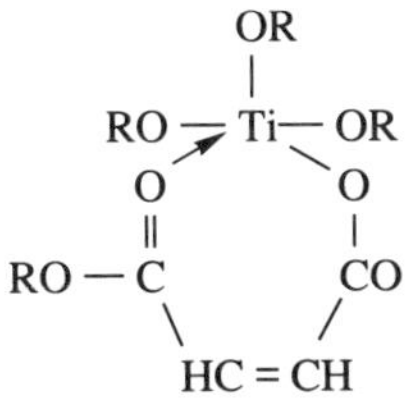

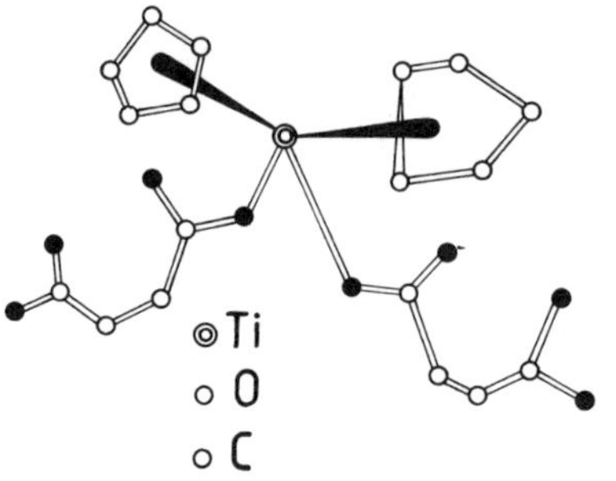

Figure 5 Crystalline structure of cyclopentadienyltitanium maleate.

Detailed consideration has been given to the interaction of maleic acid with $(C_5H_5)_2TiCl_2$. Thus, it has been possible to isolate $(C_5H_5)_2Ti(OCOCH{=}CHCOOH)_2$ in the aqueous medium.[118] The complex has an orthorhombic configuration (Figure 5), and the titanium atom a pseudo-tetrahedral one. The maleate anions are associated with the titanium atom in *cis*-conformations and are positioned in the same crystallographic plane.

Also, maleates of some other metals such as Mn(II), Co(II), Ni(II),[119], Mg(II), Ca(II), Zn(II), Cu(II),[120] and Pr(III)[121,122] have been synthesized.

Much promise is shown by those MCMs that represent derivatives of metals and an ethylene glycol methacrylic derivative of phtalic acid (ethylene glycol methacrylate-phthalate) [EMP] that are produced as follows:[123,124]

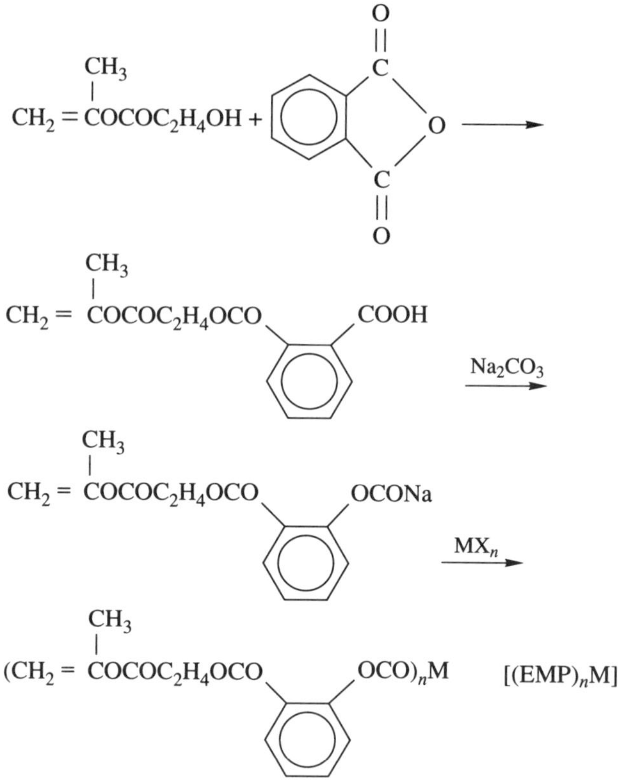

Among the derivatives of this type based on nontransition metals only a magnesium salt is soluble in ether, and a Pb(II) salt precipitates in the form of a viscous liquid. The reaction yield is 30 to 92%. In the case of Cr(III), Fe(III), Co(II), Ni(II), and Cu(II) the process will be more sophisticated. The highest yields and purity are provided by an $(EMP)_2CrOH$ salt.

II. HOMOPOLYMERIZATION OF IONIC-TYPE MCMs

A. RADICAL POLYMERIZATION OF SALTS COMPOSED OF ALKALI OR ALKALI-EARTH METALS AND UNSATURATED SALTS

Polymerization of alkali and alkali-earth metal salts based on both weak (AA, maleic) and strong (vinyl succinimidic and sulfo) unsaturated acids occurs, as a rule, in aqueous or aquo-organic acid environments at pH >7.0. In these conditions MCMs are virtually dissociated; e.g., at pH 3.0 the dissociation of MAA is no more than 3% whereas at pH 7.0 it is as high as 99%, and monomers are represented by the corresponding anions. Polymerization features of MCMs in such a state ("ionizable monomers") can be interpreted with the use of a hypothesis describing the kinetic role of ion pairs.[126] According to this hypothesis, in the alkaline pH region the rate of chain propagation is only governed by that of macroradical interaction with a terminal ion pair. The kinetic effects observed, in particular, the values of $k_p/k_t^{1/2}$, are explained in terms of changing the effective macroradical reactivity in that the propagating ionized macroradicals comprise separated ion pairs or associates. The metal cation can be thought to act as a counterion which

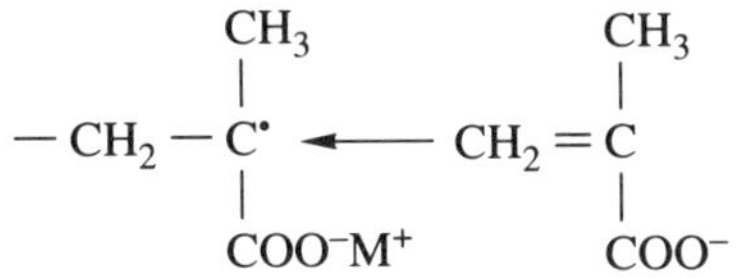

affects the ion pair stability and the chain growth rate:

We shall restrict ourselves to examining the seemingly most typical situations. Thus, sodium methacrylate polymerization in concentrated aqueous solutions (3.15 to 4.67 mol/l) in the presence of $K_2S_2O_8$ obeys the principles of zero-order reactions with respect to the monomer, the order with respect to the initiator being 0.51 and the apparent activity energy (E_a) values (between 50 and 75°C) being 91.5 kJ/mol.[127]

In the optimized process one can obtain linear water-soluble alkali metal-containing polymers with enhanced molecular mass (MM) ([η] >5 dl/g in 2 *N* aqueous NaCl solution at 25°C).[128]

Interestingly, the presence of metal in the MCM molecule does not preclude emulsion (latex) polymerization in which ordinary salts cause latex coagulation. Used as emulsifiers for sodium acrylate polymerization are both ionogenic and nonionogenic surfactants such as oxyethylated alkyl phenols.[129] Of great interest is the polymerization of MCMs incorporating surfactant groups.[103] Polymerization conversions of such MCMs can be exemplified by homopolymerization of an undecylenic acid sodium salt $CH_2{=}CH(CH_2)_8OCONa$ which in water gives rise to micelles.[130] The process initiated by ^{60}Co radiation does not start until a specified monomer concentration (at least 0.043 mol/l) is attained. It is obvious that for the polymerization to occur micelle formation is essential, and chain propagation acceleration is due to the double bond localization at the micelle surface.

The state of micelle aggregate association is likely to significantly affect the polymerization MCM properties. A typical example is a spontaneous radical polymerization

Table 5 **Some kinetic parameters and E_a of polymerization for ionic-type alkaline metal-containing monomers**

MCM	Polymerization conditions				Reaction order			
	Initiator	Medium	pH	T (°C)	with respect to monomer	with respect to initiator	E_a (kJ/mol)	Ref.
$NaOCOCH{=}CH_2$	$(NH_4)_2S_2O_8$	Water	7.2	40–60	2	1	70.1	132
$NaOCOC(CH_3){=}CH_2$	AIBN	Water	4–12	61	1	0.5	–	132
	$K_2S_2O_8$	Water	–	50–75	0	–	91.5	127
	$K_2S_2O_8$	Water-chitosan acetate	–	30–70	0.8	0	–	133
K-salt of *N*-vinyl-succine-imidic acid	H_2O_2	Water	–	40–50	2	0.7	84	132, 134
Na *p*-styrene sulfonate	$K_2S_2O_8$	Water-dioxane						
		4:1	–	70	1.04	0.5	–	135, 136
		1:1	–	70	1.01	0.5	–	135, 136
		Water	–	70	1.28	0.5	–	135, 136
	AIBN, $K_2S_2O_8$	Water-DMSO						
		1:3	–	70	1.3, 135	0.8, 0.65	42	137, 138
		3:1	–	70	1.1, 1.4	0.5, 0.82	37.6	137, 138
		Water	–	70	1.2	0.5	46.2	137
		DMSO	–	70	1.5, 15	0.95, 0.85	32.8	137, 138
Na methallyl sulfonate	γ-Radiation	Water	–	20	0.73	0.60	–	139

of sodium alkyl-2-oxy-3-methacryloxypropyl phosphate in water or benzene at 5 to 60°C.[131] In this case the polymerization monomer ability is heavily dependent on the type of the alkyl radical used and grows in the series $C_{16}H_{33} < C_{14}H_{29} < C_8H_{17} < C_{12}H_{25}$.

A well polymerizable class of ionogenic monomers is represented by salts of strong unsaturated acids (Table 5). A typical example is homopolymerization of alkali metal *p*-styrene sulfonates (H_2O-dioxane, 70°C, $K_2S_2O_8$).[135,136] Such parameters as polymerization rate (w), the reaction order with respect to the monomer, and MM of the resulting polymer decrease with dioxane content in the mixture. This is attributed to the effect of the electrostatic interaction on the composition of the propagating radical associates and MCM molecules. A similar behavior is observed in the polymerization of these MCMs in H_2O-dimethyl sulfoxide (DMSO) mixtures in that the initial rate and E_a of the process as well as MM of the polymer grow with water fraction in the mixture. As this takes place, the reaction order with respect to the initiator is far in excess of 0.5[137,138] An increase in the fraction of DMSO, a bipolar aprotic solvent, decreases the degree of MCM and macroradical salt group polymerization and stabilizes the $-SO_3^-\ldots M^+$ ionic pairs.[138] This will necessarily weaken the electrostatic interaction between the monomer and the propagating radical, which is displayed in the k_p and k_t values.

The radical polymerization of *p*-styrene sulfonates is markedly affected by the solution ionic strength generated by additives of a nonpolymerizable salt, e.g., KCl.[136, 140] The initial polymerization rate and the $k_p/k_t^{1/2}$ ratio increases with MCM and KCl concentrations (up to 0.4 and 1 mol/l, respectively). As this takes place, the reaction order with respect to the monomer also grows (from 1.2 in water to 1.5 in 1 *M* KCl) with a parallel drop of E_a from 47.9 to 26.5 kJ/mol; the reaction order with respect to the initiator remains equal to 0.5.

In addition, plasmochemical polymerization of these MCMs has been conducted.[105] To do so, the aqueous solutions of sodium sulfohexyl- and sulfoethylmethacrylates, as well as of vinyl sulfonate were plasma treated (with a 10 MHz discharge) for 5 min at –10°C with subsequent postprocess polymerization at 5 to 50°C. The polymerization activity was found to change in the MCM series sulfohexyl methacrylate > sulfoethyl methacrylate > vinyl sulfonate to suggest consistency with the data on MCM polymerization using physical-type initiators.

To a large measure, the above features apply to the polymerization of alkali-earth metal salts.[141] In the concentration range 0.12 to 0.3 mol/l the initial polymerization rate of *p*-styrene sulfonates, $k_p/k_t^{1/2}$ values, a reaction order with respect to the monomer, and the mean MM of the polymers increase in the series of cations in the ascending order of their radii: Ca(II) < Sr(II) < Ba(II), the order with respect to the initiator being equal to 0.5 and with respect to MCMs close to 1.0. As the cation radius increases, the electrostatic repulsion between the monomer and the propagating macroradical decreases to lead to a larger $k_p(k_p/k_t^{1/2})$. The influence of cation type is more pronounced at smaller ionic strengths. "The salt effect" is connected with the participation of the ion pairs in the propagation steps, the concentration of these pairs increasing as nonpolarizable salts are added to the reaction mixture.

Much attention previously has been paid to the polymerization properties of AA salts. The maximum polymerization rates—$(NH_4)_2S_2O_8$ initiation—of Mg(II), Ca(II), and Sr(II) acrylates ($w \cdot 10^6$, mol/l·s) are 160, 433, and 400, respectively; and MM values are 92,500, 848,000, and 990,000, respectively.[142] These differences are attributed to a dissimilar charge density at the macroradical anion, which influences the rate of interaction in the "propagating macroradical-monomeric anion" system.

It should be taken into account that alkali-earth metal-containing MCMs, like most transition metal-based monomers, comprise divinyl-type species with unconjugated double bonds. Such MCMs can be homopolymerized in two ways: (1) with the opening of one or both multiple bonds and (2) with the appearance of either linear or cross-linked polymers:

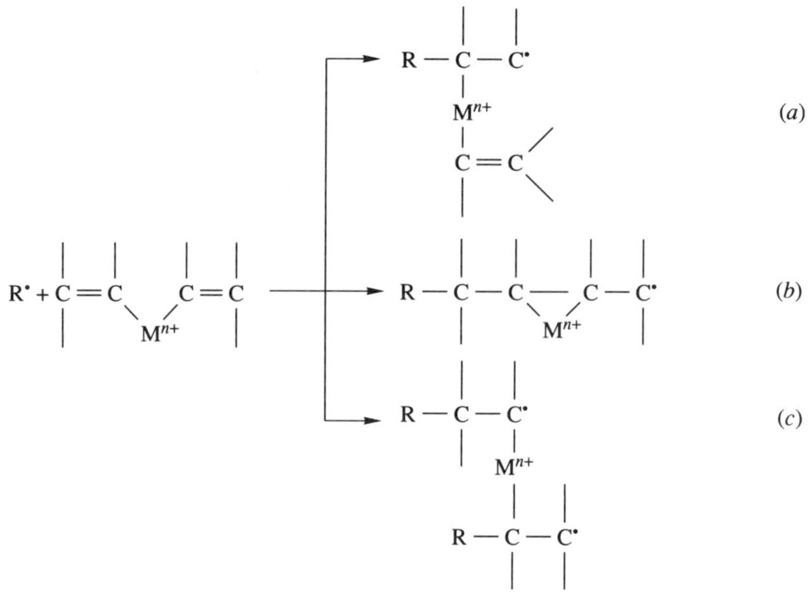

No studies illustrating the contribution of cycloaddition as exemplified by dimethacrylates have been made. However, the insolubility of the corresponding polymers may suggest a significant role of stage (*b*) leading to the emergence of structures cross-linked through a metal ion. The mechanism of cyclic unit formation seems to be close to the polymerization of unconjugated divinyl monomers (diene-type, methacrylic anhydride, etc.). It resides in the alternation of the inter- and intramolecular chain propagation reactions:

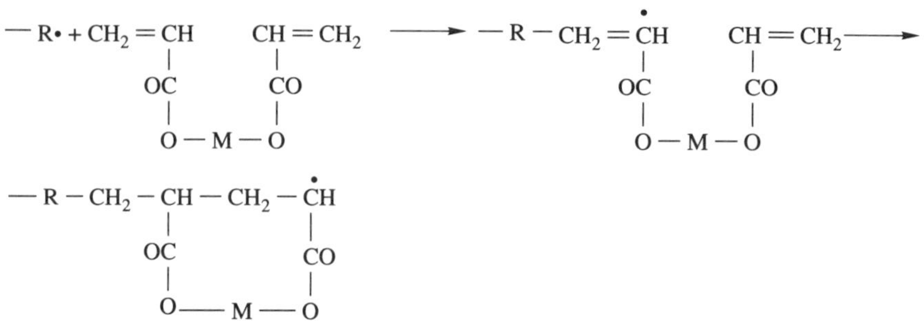

Radical polymerization of ionizable allyl-type MCMs frequency involves a decreasing role of a degradative and an increasing role of an effective chain transfer process. Thus in polymerization of sodium methallyl sulfonate in water, water-methanol mixtures, DMFA, and dioxane one can observe a deviation of the process from a typical "allyl" kinetic behavior when $w = K[M]^0[I]^{1.0}$; $w = K[M]^{0.73}[I]^{0.60}$.[139] This is due to a reinitiated chain transfer in the presence of polymeric allyl radicals, decelerated mortality of these radicals, or their increased reactivity, etc. At the same time, the resulting polymers have a short chain ([η] = 0.027). Then, sodium allyl sulfate polymerization in pure DMSO or its aqueous solution is distinguished by ordinary features typical of "allyl" polymerization, despite the fact that depending on solvent type the monomer is represented by various types of moieties such as $CH_2{=}CH{-}CH_2SO_3^-$ ...Na^+ (in DMSO), $CH_2{=}CH{-}CH_2SO_3^-$ (in water) or $CH_2{=}CH{-}CH_2SO_3H$ (in DMSO at pH 1.5).

B. RADICAL POLYMERIZATION OF TIN ACRYLATES

Early polymerization studies of tin- and lead-containing acrylates are reviewed elsewhere.[9]

We shall restrict ourselves to the analysis of major kinetic homopolymerization parameters as exemplified by the most recognized tin acrylates such as tributyltin methacrylate (TBTM) and tributyltin acrylate (TBTA).

Depending on the initiation efficiency of radical TBTM in-bulk polymerization (AIBN initiation, 60°C), the $k_p/k_t^{1/2}$ ratio can be very large (0.38 to 0.57), which is well in excess of that in alkyl methacrylate polymerization. At high monomer concentrations the reaction order with respect to the initiator becomes less than 0.5. Such a kinetic behavior can be attributed to termination at primary radicals. An increase in the reaction order with temperature (from 0.4 to 0.5), when this type of termination predominates at lowered temperatures, is due to the fact that the E_a value of the viscous monomer flow is high (17.6 kJ/mol) and is commensurate with that of termination; whereas the E_a value of propagation (19.3 kJ/mol) is commensurate with that in MAA polymerization. In other words, a higher polymerization rate for TBTM as compared with that for MAA is caused by labored reaction termination as a result of steric hindrances and high MCM viscosity.

Homopolymerization of TBTA and TBTM in the presence of solvents is characterized by formation of low molecular products only with degree of polymerization (P_n) < 40.[44,45] This might be due to chain transfer to the solvent, e.g., in mixtures composed of TBTA and decane-benzene and of TBTM and *m*-xylol (butyl acetate). The kinetic equations of the processes are close to the classical forms: $w = [I]^n[M]^m$, where $n = 0.45$ to 0.49 and $m = 1.05$ to 1.10. Interestingly, P_n of the two MCMs (azobisisobutyronitrile [AIBN] initiation) is virtually independent of the reaction conditions (within 8 to 10), which appears due to a pronounced chain transfer to the initiator.

A low polymerization activity of tin acrylate monomers in the presence of solvents is additionally corroborated by data on the polymerization of *n*-acryloxytributyltin benzoate

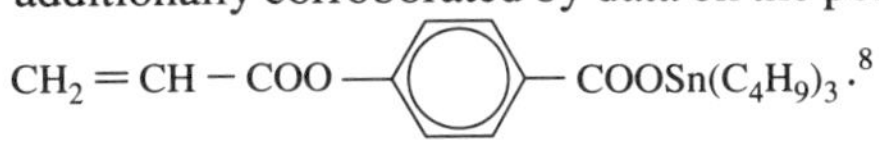

.[8] The product yield in 24 h at 60°C in DMFA (AIBN initiation) is as low as 15%.

C. RADICAL POLYMERIZATION OF TRANSITION-METAL METHACRYLATES

There are reports that Zn(II), Co(II), Ni(II), and Cu(II) acrylates polymerize at 60°C (AIBN initiation) in dissociation-excluding conditions.[83] As this takes place, the Zn(II) acrylate polymerization rate is maximum, although the process occurs following a significant (around 100 min) inductive period. For the remaining acrylates no inductive period is observed whereas their polymerization activity is markedly different. For example, the rate of Cu(II) polymerization is very low. The resultant metal-containing polymers are insoluble in organic solvents but they convert to PAA in a methanol-HCl mixture. This approach is used as an indirect method for determining the MM values of the products.

In comparable conditions the rate of transition metal acrylate polymerization (AIBN initiation, in ethanol) decreases as follows: Co(II) > Ni(II) > Fe(III) > Cu(II), and in all cases it is lower than that of AA polymerization (Figure 6).[146] A special nature of Zn(II) acrylate polymerization properties has been corroborated when studying the polymerization of this MCM in DMFA (AIBN initiation, 65 to 75°C).[147] The associated kinetic equation is as follows: $w = K[I]^{0.86}[M]$.[149] At the same time, the kinetic equations of Co(II) and Ni(II) acrylate polymerization in similar conditions resemble those that are typical of radical polymerization with bimolecular termination:[148]

$$w = K[M]^{1.27}[I]^{0.54} \text{ for Co(II) acrylate}$$

$$w = K[M]^{1.2}[I]^{0.53} \text{ for Ni(II) acrylate}$$

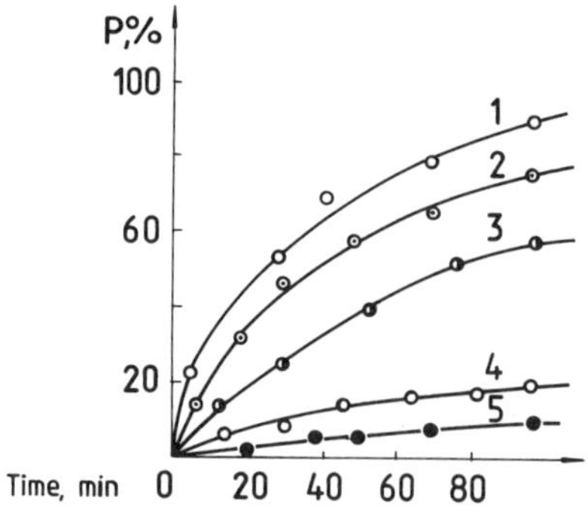

Figure 6 Kinetic curves of homopolymerization yield for AA (1); and metal acrylates: Co(II) (2), Ni(II) (3), Fe(III) (4), and Cu(II) (5). [MCM] = 0.9 mol/l. [AIBN] = $2.5 \cdot 10^{-2}$ mol/l, ethanol, 78°C.

It is interesting that the above order of MCM activity changes for alcohol environments also remains the same in DMFA, namely, Zn(II) > Co(II) > Ni(II). However, the E_a values (in kJ/mol) of this process change in the reverse order: 89.8 > 74.0 > 71.6 for Ni(II), Co(II), and Zn(II), respectively.

Titanocene mono- and dimethacrylates are fairly well polymerizable (80°C, DMFA, 0.3 to 0.5 wt.% benzoyl peroxide [BP]) to give the following polymeric structures:[96]

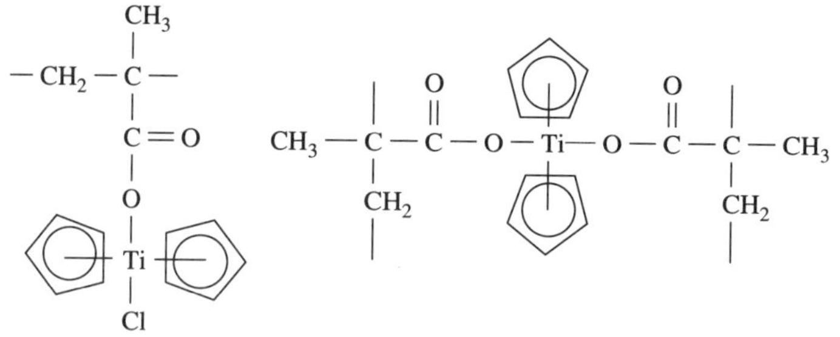

The former is soluble in acetone, DMFA, and dioxane, and the latter is insoluble in organic solvents. Another Ti(IV)-containing monomer, tributoxytitanium methacrylate, does not polymerize in the presence of BP and on heating to 80 to 100°C.[149] This is attributed to the BP-alkyl orthotitanate interaction to give rise to titanium acrylate, with which the reaction is noncompetitive.[150] Nevertheless, in the presence of 2% AIBN on heating to 150°C, it has been possible to polymerize this MCM. At the same time, dibutoxytitanium maleate polymerizes spontaneously even at the stage of synthesis to afford a product soluble in many organic solvents (MM = 2230).[90]

A fairly detailed consideration has been given to the polymerization properties of titanium-containing macromers, i.e., oligomers of the type:

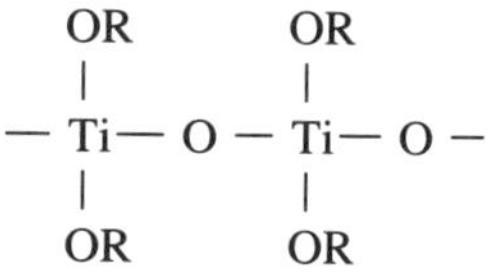

where part of the framing groups contains AA, MAA, furyl-acrylic, or cinnamic acid residues.[91-93] The behavior of polybutoxytitanoxane methacrylate is similar to that of a low molecular analogue, tributoxytitanium methacrylate, in that in the presence of BP one can observe no polymerization even in the case of prolonged heating; whereas in the presence of AIBN and on heating for 5 h at 150°C, the number of double bonds significantly decreases.

Kinetic analysis of transition metal acrylate homopolymerization indicates that this process is characterized by the same elementary steps as with ordinary vinyl-type monomers, although sometimes it is complicated by a purely individual influence of the transition metal.[146] The canonical scheme of radical polymerization also can be used to describe the polymerization of these MCMs:

$$I \xrightarrow{k_i} 2R^{\bullet},$$

$$R^{\bullet} + M \xrightarrow{k'_p} R - M^{\bullet}$$

$$R - M^{\bullet} + M \xrightarrow{k_p} R_j^{\bullet}$$

$$R_j^{\bullet} + R_s^{\bullet} \xrightarrow{k_t} P,$$

where $R^{\cdot}$, $R—M^{\cdot}$, $R^{\cdot}_{j,}$ and $R^{\cdot}_s$ are the initiating, primary, and propagating polymeric radicals, with the corresponding number of monomeric units, respectively; P is the resulting product.

Considering the fact that the initial $[M]_0$ and the current monomer (M) concentrations are related to the conversion α by the relationship $[M] = [M]_0\ (1 - \alpha)$ and a quasi-stationary approximation with respect to the macroradicals is fulfilled, the polymerization rate can be expressed as follows:

$$d\alpha/dt = k_p(k_i/k_t)^{1/2}I^{1/2}(1 - \alpha).$$

Given that $I = I_0^{e - k_i t}$, its solution leads to the following equation:

$$\ln\{\ln\,(1 - \alpha) + 2k_p[I_0/k_i,k_t)]^{1/2}\} = \ln\,2k_p[I_0/(k_i \cdot k_t)]^{1/2} - 1/2k_i t,$$

which provides a good description of the polymer buildup kinetics. Also, such an analysis characterizes the observed dependence of the limiting conversion on initial initiator concentration:

$$\ln(1 - \alpha_{\infty}) = 2k_p[I_0/(k_p \cdot k_t)]^{1/2}.$$

The k_i value as found from these equations is $3.16 \cdot 10^{-2}$ min which is close to the published data.[151]

A correlation between the rate of transition metal acrylate polymerization and metal electronegativity has been established. With increasing electronegativity the electronic double bond density drops with the result that the rate of the process decreases.[152] However, the reasons for the effect produced by one or another transition metal on the polymerization kinetics of such MCMs are more profound and, to a large extent, individual. Therefore one can mention, primarily, a low polymerization activity of Cu(II)- and Fe(III)-containing MCMs, which is likely to result from an intramolecular chain termination:

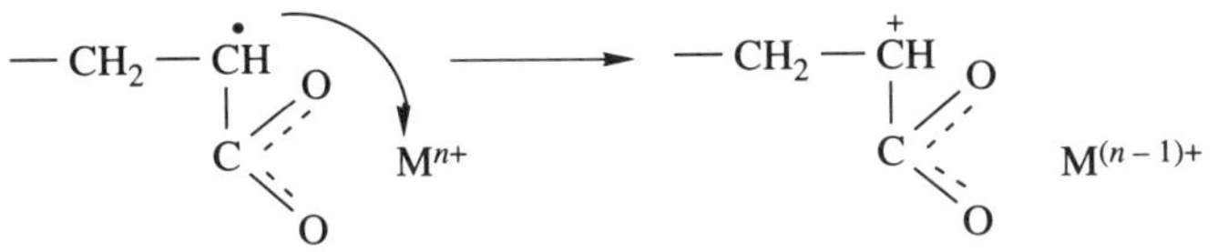

As this takes place, the electron of the propagating macroradical goes to the metal ion and decreases the degree of its oxidation. This process is favored by fairly high standard electrode potentials of the copper and iron ions:

$$E^0_{Cu^{2+} \rightarrow Cu^{+}} = 0.15 \text{ V}$$

$$E^0_{Fe^{3+} \rightarrow Fe^{2+}} = 0.77 \text{ V}$$

Polymerization of the MCMs under study in the aqueous environment is complicated by dissociation processes. In particular, in contrast to radical polymerization of alkali and alkali-earth metal methacrylates, homopolymerization of a methacrylic derivative of pentaamminocobalt(III) perchlorate ($K_2S_2O_8$—$Na_2S_2O_3$ initiation) occurs at pH 3.0 when the MCM is not dissociated; whereas at pH 7.0, no reaction takes place. This fact is attributed to the electrostatic repulsion of the like charges of the propagating radical and of the monomer introduced.[88]

D. POLYMERIZATION OF IONIC-TYPE MCMs BASED ON LANTHANIDES AND ACTINIDES

This branch of MCM polymerization is still at the embryonic stage, but its promising future justifies these monomers to be placed in a separate section.

Homopolymerization of La(III), Eu(III), Pr(III), and Y(III) acrylates have been reported.[99] Unusual polymerization properties of Pr(III) maleate (H_2O_2 initiation) have been detected.[121,122] The role of hydrogen peroxide has been found to consist of Pr(III) oxidation to Pr(IV). The Pr(IV) ion with unoccupied orbitals at the outer electronic level seems to form complexes with the maleate ion and water that initiate polymerization. One of these complexes is assumed to comprise a pair of H^+ $[Pr(C_4H_2O_4)_2OH]^-$ ions. The hydrogen ion of the complex connects at the double monomer bond to give a carbonium which in turn takes up the next MCM molecules to afford the macromolecule:

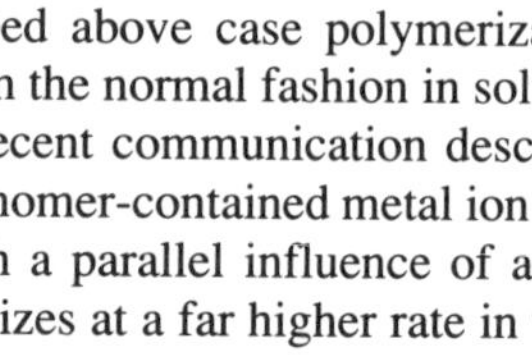

As distinct from the described above case polymerization of a methacrylic uranium dichloride derivative occurs in the normal fashion in solution (AIBN initiation).[100] There is a remarkably interesting recent communication describing the possible effect of the isotopic composition of a monomer-contained metal ion on polymerization properties. In the alcohol environment with a parallel influence of a high-frequency magnetic field, uranyl acrylate photopolymerizes at a far higher rate in the case of ^{235}U.[101] Interestingly, in the absence of a magnetic field the process does not take place at all, which is attributed to the peculiar magnetic properties including the presence of a magnetic dipole at ^{235}U. Note that these data could not be reproduced in a recent work.[190]

E. LOW-TEMPERATURE RADICAL MCM POLYMERIZATION

One of the problems in MCM polymerization at lowered temperatures is the existence of a limited range of free-radical polymerization initiators. The fact is that the conventional types of initiators are only suitable for being used at elevated temperatures, and redox systems fairly frequently give rise to undesirable side effects. Nevertheless, recently there have appeared organometallic compounds that can be used as radical initiators. Thus, the cationic Co(III) complexes

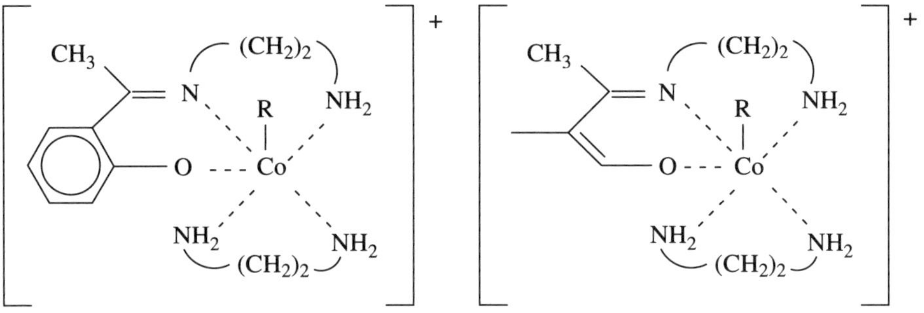

(R = CH_3, C_2H_5, or C_3H_7) undergo homolytic decomposition at the Co—R bond.[153] Therefore, they have been used for low-temperature polymerization of such MCMs as Zn(II), Mg(II), Pb(II), and Ba(II) in methanol.[154] The general scheme of the process considering a partial monomer dissociation and participation of resultant cations in decomposition of the complex will be as follows:

$$R-Co \xrightarrow{[M(OCOCH=CH_2)]^+} R^\bullet$$

$$\underset{\displaystyle COOM(OCOCH=CH_2)}{CH_2=\underset{|}{CH}} + R^\bullet \longrightarrow -CH_2-\underset{\displaystyle COOM(OCOCH=CH_2)}{\underset{|}{CH}}-$$

The kinetic behavior of the polymerization process and its complete inhibition in the presence of compounds that act as radical traps provide support for the radical nature of the process. Such initiators allow the process at relatively low temperatures and fairly high rates (Figure 7).

Another possible way for obviating the side reactions is through radiation initiation which makes the process possible at any phase state of the monomer in a wide temperature range. In recent years there has been an extensive application of a postradiation polymerization method with matrix devitrification, i.e., as the system changes to a supercooled liquid.[155] Thus on heating 10% alcohol solutions of transition metal acrylates, the colorimeter indicates a typical heat capacity change due to devitrification.[156] On heating a glassy Ni(II) acrylate solution irradiated at –196°C immediately after the devitrification "step" in the temperature range between –170 and –138°C one can observe heat release associated with the polymerization of this MCM. The polymer forms a separate phase; and the colorimetric curve shows the presence, along with an exothermic

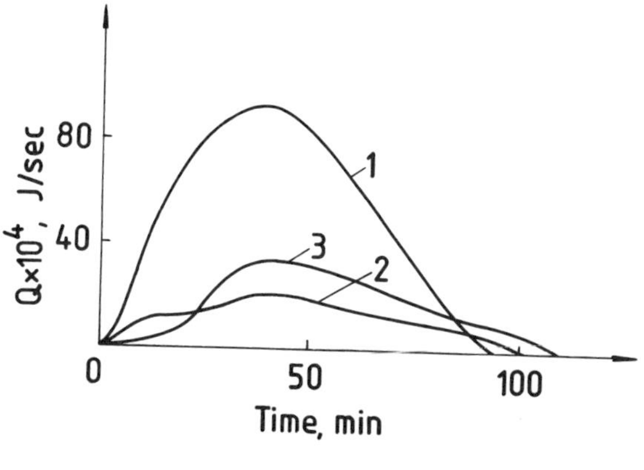

Figure 7 Heat release during low-temperature polymerization of metal acrylates: Zn(II) (1), Ba(II) (2), and Pb(II) (3). Heating rate = 0.2 K/min.

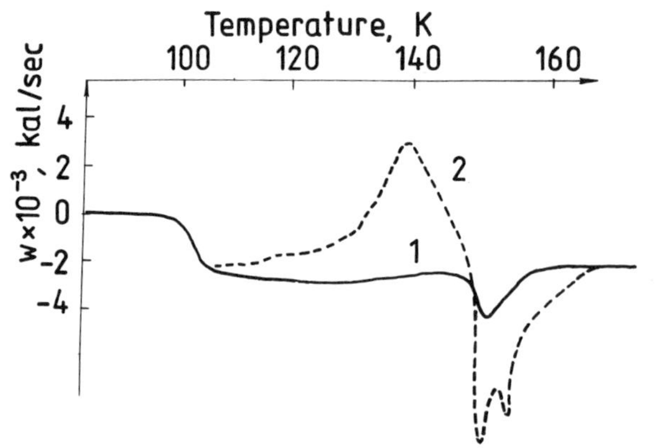

Figure 8 Calorimetric curves for heating 10% Ni(II) acrylate solution in ethanol prior to (1) and following (2) irradiation at 100 kJ/kg and −196°C.

peak of polymerization, an exothermic peak of crystallization and an endothermic peak of melting (Figure 8). These facts are associated with increasing molecular reagent mobility in the phase transfer region and with intermolecular interaction in the alcohol matrix. In such conditions the heat of the Co(II) acrylate polymerization is 114 kJ/mol.

F. POLYMERIZATION OF IONIC-TYPE MCMs IN THE SOLID PHASE

As shown above, MCM polymerization in solutions has some disadvantages such as dissociation of salts, especially of alkali and alkali-earth metal ones, as well as a limited number of solvents which are suitable for producing fairly concentrated MCM solutions. At the same time, most MCMs of this type are solid substances at room temperature, so that solid-phase polymerization methods can be applied to them. In addition, these MCMs are convenient for being polymerized using these methods considering their chemical and structural features. The orientation of their molecules is such that often they can be conveniently interconnected through chemical bonds. In this case the chains grow in the plane of the already available "blanks" or stacks in the form of parallelly disposed and closely packed MCM molecules. Therefore, for the process to take place, no significant change in the position of the molecules as compared with that in the parent crystal is required.

Irrespective of how the solid-phase polymerization is initiated, its kinetic scheme is based on the following assumptions: the spatial movement of the growing macroradicals and their collision with the monomer molecules are only due to the chain growth steps (as a result of a nearly complete lack of the reacting species diffusion); crystal lattice inhomogeneities (dislocations, cracks, vacancies, etc.) are places where the growing chains terminate; the reactivity anisotropy of the macromolecules growing in the crystal lattice is responsible for their predominant propagation along one of the crystallographic axes.

The process of excitation of radical (less frequently ionic) polymerization in the solid state is implemented by using various types of initiation such as thermal, photochemical, radiation, and sometimes mechanical.

G. THERMAL POLYMERIZATION

The thermal generation of free radicals in the polymerization of the MCMs under discussion can be exemplified by rare cases and relate mostly to nontransition metal acrylates. Polymerization-initiating radicals emerge as a result of a C—C or C—H bond dissociation or double bond opening (biradicals). For the reactive centers to come closer to one another, the thermal oscillation amplitude should be high enough; therefore, the solid-phase polymerization becomes more pronounced with temperature. Thus, the thermal polymerization of sodium acrylates takes place at 145 to 175°C *in vacuo*.[157] The E_a value of initiation as determined from nuclear magnetic resonance (NMR) evidence is equal to 121 kJ/mol. The process involves an induction period which decreases with temperature. The E_a value found from these data is 116 kJ/mol, which is consistent with the above value. The chain propagation stage is characterized by E_a = 68.9 kJ/mol.

When sodium acrylate is heated in the presence of atmospheric oxygen, the thermal polymerization temperature is much higher (270°C).[65] The polymerization temperature of alkali-earth metal acrylates is somewhat lower: ca. 230°C for Mg(II) and Zn(II) and ca. 180°C for Ca(II) species. The E_a value of such processes does not exceed 33.6 kJ/mol. The oxidation decomposition of the polymer takes place at 400 to 550°C.

One of the first MCMs that was polymerized in the solid phase was $Ba[OCOC(CH_3){=}CH_2]_2 \cdot H_2O$.[20] In the dehydration of this MCM at 65 to 140°C, the following general reaction takes place:

$$Ba(OCOC(CH_3){=}CH_2)_2 \cdot H_2O \rightarrow R^{\bullet} + H + CO_2 + HO^{\bullet}$$

with the appearance of various radical-type species which can initiate polymerization.[70,158] In the temperature region 190 to 350°C, Ba(II) and Ca(II) acrylates give rise to di-, tri-, and tetramers.[159]

Transition metal acrylates also can be polymerized[160] at the following temperatures (°C): 180 to 265 for Zn(II), 230 to 320 for Co(II), 250 to 320 for Ni(II), and 205 to 242 for Cu(II) varieties. At higher temperatures the polyacrylates decompose. Interestingly, both radical initiation in methanol[83] and solid-state thermal polymerization exhibit the same features in terms of reactivity: high for the Zn(II), somewhat lower for Co(II) and Ni(II), and the lowest for Cu(II) acrylates.

H. SOLID-PHASE ULTRAVIOLET AND RADIATION-INDUCED POLYMERIZATIONS

Photochemical initiation of MCM polymerization in the solid phase is a relatively rare case. One such example is an effective potassium acrylate polymerization.[161] In the case of ultraviolet (UV) radiation at 250 to 300 nm, the polymer yield is as high as 40% at 73°C but its MM decreases from $4.7 \cdot 10^5$ to $2.2 \cdot 10^5$ as conversion increases. The photoinitiated Cu(II) acrylate polymerization also occurs at a constant rate up to deep conversions.

A solid-phase polymerization in the multilayers of Ag(I) diacetylate $CH_3{-}(CH_2)_{17}{-}C{\equiv}C{-}C{\equiv}C{-}COOAg$ (UV initiation at 252 nm) proved to be fairly efficient.[107] The reaction proceeds with 50% yield and virtually does not occur in the case of other metal salts.

A radiation-induced polymerization is a universal method for solid-phase systems. As the MCMs are exposed to ^{60}Co γ-radiation, accelerated electrons or X-rays at low temperatures (usually –196 or –78°C), there will be accumulation of radicals in the solid phase for subsequent postpolymerization at a higher temperature.

Data on the ready polymerization of potassium and rubidium acrylates *in vacuo* at room temperature (irradiation at –78°C) have been reported even in early works, whereas lithium and sodium salts should be heated up to 150°C.[56] The most reactive MCM is potassium acrylate: at 0°C the rate of its polymerization (E_a = 70 kJ/mol) is higher than that of sodium salt at 120°C. The chain length of the potassium polyacrylate chain is one order of magnitude greater than that of sodium or lithium. It is interesting that for MAA salts the reverse behavior has been observed in that sodium methacrylate is more active than the potassium one, whereas lithium salt is inactive. These differences are related to the crystal lattice geometry of the corresponding salts, which in turn is governed by the type of the metal ion. The minimum rate constant for radical mortality in irradiated potassium and rubidium acrylates ($0.23 \cdot 10^{-3}$ and $2.28 \cdot 10^{-3}$/s, respectively) are observed at temperatures approximating those of phase transitions of these MCMs, at 61 and 47°C, respectively.[59]

Some interesting features are typical of Ca(II) acrylate polymerization in that the polymer yield displays a significant dependence on the degree of MCM hydration. On irradiation at –78°C, when conducting the postprocess at 25°C, the maximum yield is attained when a semihydrate form of polymer is employed.[61,67] According to other data, the maximum polymerization rate for this MCM is observed in the case of acrylate monohydrate.[69] Interestingly, the polymerization behavior of the anhydrous Ca(II) acrylate depends heavily on how it is obtained (dehydrated), i.e., from a di- or monohydrate. In the first case the reactivity is higher.

Also, Ba(II) methacrylate polymerization depends significantly on the presence of crystal water.[61,62,68,69] Note that even after high-dose irradiation and prolonged heating the anhydrous MCM gives at least 2% of polymer. The maximum rate of the process is observed at a water content of 0.25 mol/mol salt. It is obvious that at such water contents the MCM structure is the loosest and contains a lot of dislocations facilitating the solid-phase polymerization. The hydrogen atoms appearing from the crystal water during salt hydrate radiolysis contribute appreciably to the formation of free radicals which initiate the solid-phase polymerization. The electron spin resonance (ESR) spectrum of the γ-irradiated Ba(II) acrylate dehydrate is assigned to the radicals which are formed as follows:[162,163]

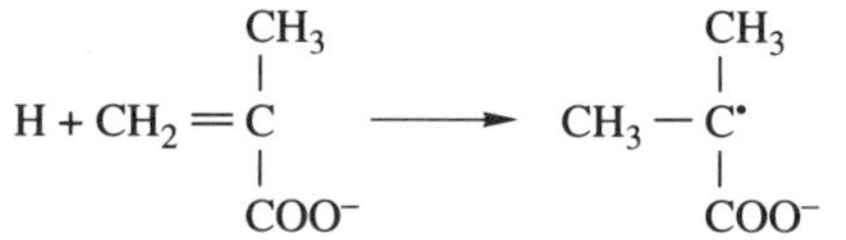

The content of such radicals is as high as 90%; the balance is the growing species of the type $RCH_2C(CH_3)COO^-$. Experiments with D_2O indicate that around 75% of the initiating radicals in the monohydrate of this MCM are formed by adding crystal water hydrogen atoms to the double bond. In this case the polymer yields are 4.7 1/100 eV and 8.8 1/100 eV at –196 and –78°C, respectively;[164,165] $w \sim I^{0.5}$; E_a = 132 kJ/mol. It should be noted that in Ba(II) acrylate dihydrate polymerization, the conformation of the growing radicals differs from that of the monohydrate. According to X-ray structure analysis, the crystal structure of $Ba(OCOCH{=}CH_2)_2{\cdot}2H_2O$ is significantly disturbed in the course of polymerization.[71] The induction polymerization period observed is associated with the capture of the growing short-chain radicals, and its duration markedly decreases with either temperature or irradiation dose.

In contrast to Ba(II) and Ca(II)-containing polymers, the anhydrous Zn(II) acrylate displays a clear-cut tendency to radiation-induced postpolymerization (irradiation at –78°C, polymerization at +89.5°C).[157]

Among the limited information on the γ-initiation of solid-phase transition metal acrylate polymerization we shall note a high tendency of a Fe(III) methacrylate toward polymerization. Its 90% conversion to polymer at 25°C takes no more than 8 h.[166]

I. SOLID-PHASE POLYMERIZATION UNDER PRESSURE, HIGH-PRESSURE/SHEAR STRAIN OR MECHANOCHEMICAL INITIATION CONDITIONS

Radicals capable of initiating MCM polymerization also can be obtained as a result of their mechanical initiation. Thus, a pressure as high as 650 MPa in solid-phase γ-induced polymerization of a crystalline Ca(II) acrylate dihydrate (a radiation dose of 3.3 to 6.6 J/kg at 19°C) provides a significant accelerating effect on the polymerization rate in that the conversion increases from 3 to 81%.[167] In the case of anhydrous Ba(II) methacrylate, the corresponding effect is less pronounced.

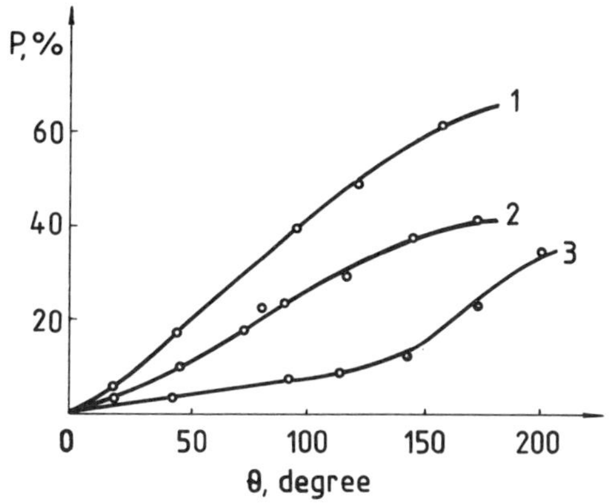

Figure 9 Polymer yield as a function of anvil turn angle under high pressure in combination with shear strains. P = 160 MPa; MCM-acrylates: Ba(II) (1), Pb(II) (2), and Zn(II) (3).

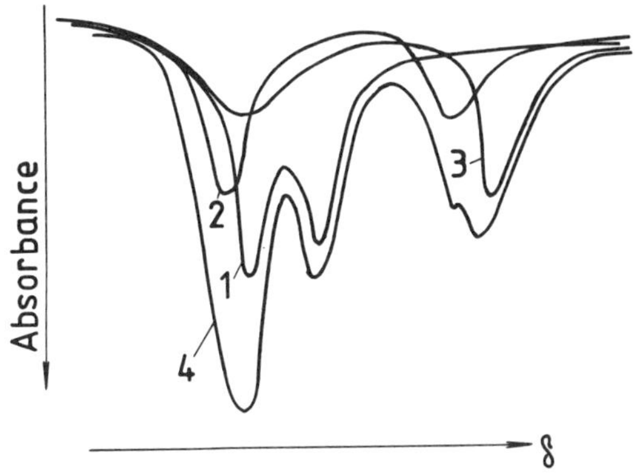

Figure 10 Mössbauer spectra for Fe(III) acrylate: prior to treatment (1), doublets as a result of treatment (2, 3), and integral after treatment (4). P = 2 GPa, anvil turn angle = 600°.

Under high-pressure conditions coupled with shear strain the MCMs represent convenient objects of study. This method implemented with a Bridgman anvil-type apparatus is very efficient for nontransition metal acrylate polymerizations (Figure 9).[168] The reactivity of Mg(II), Ca(II), Ba(II), and Pb(II) acrylates is such that high conversions are attainable even at 500 to 1000 MPa and anvil-turning angles of 100 to 300°. The addition of radical polymerization inhibitors such as benzoquinone and α-naphthol in Ba(II) acrylate polymerization enables one to determine the *w* value (8.10^{-3} mol/kg·rad).[169]

Concerning transition metal acrylates, studies of their polymerization in similar conditions have been useful in revealing a high probability of a metal ion participating in this process. Thus, the treatment of Fe(III) acrylate at 2 GPa gives rise to two more doublets in the Mössbauer spectrum (δ = 1.35 ppm, Δ = 2.11 ppm and δ = 1.44 ppm, Δ = 2.62 ppm, respectively) in addition to the initially existing Fe(III) doublet (a shift with respect to SnO_2 δ = 0.76 ppm, splitting Δ = 0.65 ppm) (Figure 10).[170] A comparison of the Fe(III) and Fe(II) spectra indicates that the new lines in the Fe(III) acrylate spectra, appearing as a result of pressure treatment, correspond to the Fe(II) acrylate. This suggests a partial reduction of Fe(III) to Fe(II) in the course of polymerization.

III. COPOLYMERIZATION OF IONIC-TYPE MCMs

A. MAJOR FEATURES OF ALKALI- AND ALKALINE-EARTH SALT POLYMERIZATION

Copolymerization of ionic-type MCMs alkali-, and alkaline-earth metal salts with various comonomers has been studied very thoroughly. What is more, it would be impossible to completely review even major works in this field.

Most studies report on a significant effect of the type of natural medium, ionic strength, and solvent polarity on the kinetic and copolymerization parameters as well as on properties of the copolymers, namely, their composition and MM. In this case the major role is attached to changes in the interaction behavior in the "macroradical-monomer counterion" systems as the medium composition is varied. Of paramount importance is the state of a given MCM (dissociated or undissociated species). In particular, the copolymerization of acrylates with acrylamide (AAm) illustrates the fact that at pH 7.1 to 7.2 when the MCM is mainly dissociated, the polymerization rate decreases in the order Na > K > Li; this reflects a lower degree of cation binding with the polyacrylamide part of the macroradical.[58] The rate of cation binding to the radicals containing acrylate groups depends on cation volume and increases in the order Li > Na > K. In other words, complex electrostatic interactions take place in the coil. The charge density in the coil will in turn affect the rate at which an undissociated MCM adds to the active site. Cation type exerts a greater influence on r_1 than on r_2 (M_2 = MCM). Hence, the effect produced by the total number of charges across the coil on r_2 is greater than that across the terminal unit. Moreover, the macroradical recombination rate is also governed by electrostatic factors. Therefore, the above series of the copolymerization rates reflects the net effect of all the known and many other factors. Some data on the copolymerization of such MCMs with conventional monomers are tabulated in Table 6.

The electrostatic repulsion of the functional salt groups and the MCM radical have been used to account for kinetic behavior in the copolymerization of MMA and alkali metal methacrylates in methanol at 60°C. In this case the reaction rate decreases with cation charge, as opposed to the situation taking place in the homopolymerization of these MCMs.[172]

Also, conclusive evidence has been given on the effect produced by the reaction environment on acrylonitrile (AN) copolymerization with sodium allyl sulfonate in DMSO-water mixtures at various pH (45°C, AIBN initiation, Table 7).[139] A change in the relative comonomer activity in various conditions is linked with their dissimilar solvating capacity. As a matter of fact, three different types of monomers take part in the copolymerization process in these systems, namely, $CH_2{=}CH{-}CH_2SO_3^-...Na^+$ (in DMSO), $CH_2{=}CH{-}CH_2{-}SO_3^-$ (in aqueous DMSO), and $CH_2{=}CH{-}CH_2SO_3H$ (in DMSO at pH 1.5). As could be expected, the parameter Q_2 is virtually independent of medium type, and the total polymerization rate decreases with MCM fraction in the monomeric mixture.

In this context, one more nontrivial fact characterizing the copolymerization of these MCMs should be mentioned. In many situations the relative reactivity constants depend on the initial composition of the monomeric mixture. Such is the case, for example, in the AN-sodium allyl sulfonate system in DMFA (50°C, AIBN initiation).[173] The Mayo-Lewis equation cannot be applied to the description of sodium styrene sulfonate-AAm copolymerization, since the copolymerization constants depend on the relationship between the initial monomer concentration and the conversion; and the resulting products are distinguished by high composition inhomogeneity.[136] Changes in the relative monomer activity also have been reported for a sodium methacrylate-methacrylamide system.[174] In addition, a marked influence is exerted by the type of comonomer. For example, a lithium methacrylate-MMA system is characterized by a strictly alternating copolymerization mechanism ($r_1 \cdot r_2 = 0.044$);[172] whereas in the copolymerization with styrene, $r_1 \cdot r_2 = 0.936$ to indicate practically complete absence of a tendency to alternation.[575]

Among other copolymerizations of ionic-type alkali metal-containing MCMs for which the same principles are observed, mention should be made of the copolymerization of sodium *p*-styrene carboxylate with styrene,[175] a random copolymerization of sodium 2-sulfoethyl methacrylate and sodium acrylamido-2-methylpropane sulfonate with AAm (aqueous solutions, 30°C, sodium persulfate initiation),[176] *N,N*-methylene bisacrylamide with sodium acrylate (H_2O_2-ascorbic acid initiation), potassium salt of *N*-vinylamidosuccinic acid with MMA in DMSO,[178] etc.

Table 6 **Copolymerization parameters of some ionic-type MCMs (alkali- and alkaline metal salts)**

MCM(M_2)	Parameter	Comonomer MMA	AAm	AA	MAA	Styrene	Ref.
p-$CH_2CHC_6H_4SO_3Na$	r_1	–	0.05	0.05	0.62	–	171
	r_2	–	0.7	0.8	0.75	–	
$CH_2{=}CHCOOH$[a]	r_1	–	0.60	–	–	–	58
	r_2	–	1.43	–	–	–	
$CH_2{=}CHOCOLi$	r_1	–	0.66	–	–	–	58
	r_2	–	0.37	–	–	–	
$CH_2{=}CHOCONa$	r_1	–	0.94	–	–	–	58
	r_2	–	0.30	–	–	–	
$CH_2{=}CHOCOK$	r_1	–	1.45	–	–	–	58
	r_2	–	0.24	–	–	–	
$CH_2{=}C(CH_3)OCOLi$	r_1	0.59	–	–	–	–	172
	r_2	0.073	–	–	–	–	
	Q_2	0.64	–	–	–	–	
	e_2	0.3	–	–	–	–	
$CH_2{=}C(CH_3)OCONa$	r_1	3.97	–	–	–	–	172
	r_2	0.426	–	–	–	–	
	Q_2	1.36	–	–	–	–	
	e_2	–1.18	–	–	–	–	
$CH_2{=}C(CH_3)OCOK$	r_1	5.65	–	–	–	–	172
	r_2	0.143	–	–	–	–	
	Q_2	0.54	–	–	–	–	
	e_2	0.01	–	–	–	–	
$(CH_2{=}CHOCO)_2Mg$	r_1	–	–	–	–	5.31	63
	r_2	–	–	–	–	0.18	
	Q_2	–	–	–	–	0.16	
	e_2	–	–	–	–	–0.59	
$(CH_2{=}CHOCO)_2Ca$	r_1	–	1.29	–	–	6.10	60, 63
	r_2	–	0.09	–	–	0.12	
	Q_2	–	–	–	–	0.11	
	e_2	–	–	–	–	–0.24	
$(CH_2{=}CHOCO)_2Sr$	r_1	–	2.75	–	–	4.12	60, 63
	r_2	–	0.05	–	–	0.14	
	Q_2	–	–	–	–	0.13	
	e_2	–	–	–	–	–0.06	
$(CH_2{=}CHOCO)_2Ba$	r_1	–	1.01	–	–	3.95	60, 63
	r_2	–	0.33	–	–	0.11	
	Q_2	–	–	–	–	0.12	
	e_2	–	–	–	–	0.11	

[a]Data for comparison purposes.

Table 7 **Copolymerization parameters of the AN-allyl sulfonate (M_2) system**[139]

Reaction medium	r_1	r_2	Q_2	e_2
DMSO (pH 7)	1.00	0.38	0.19	0.22
DMSO-water (6%)	1.25	0.28	0.14	0.18
DMSO (pH 1.5)	1.85	0.43	0.18	0.70

The total rate of radical alkaline-earth metal acrylate polymerization with AAm in aqueous solutions and the relation between the monomer activity constants change depending on cation type as follows: Ba(II) < Ca(II) < Sr(II), where r_2 decreases in a different order: Ba(II) > Ca(II) ≈ Sr(II).[60,179] The reaction rate drops with MCM concentration in the monomer mixture, the decreasing effect of Ba(II) acrylate being higher than that of Ca(II) and Sr(II) ones. As in the case of alkali metal acrylates, this effect is attributed to the electrochemical interaction between the growing macroradicals and acrylate anions, which depends appreciably on the specific interaction of Ca(II), Sr(II), and Ba(II) with the polyacrylamide moiety in the macroradical.

A different order of activities is observed in the copolymerization of alkali-earth metal acrylates in DMSO.[63] In these conditions K_{diss} of the MCM does not exceed 10^{-6} since the dissociation can be ignored. The Q_2 and r_2 values were insensitive to the position of the metal in the Periodic Table and are close to each other, whereas e_2 changes in the series of acrylates of the following metals: Mg(II) < Ca(II) < Sr(II) < Ba(II). The same changes are typical of the value of $r_1 \cdot r_2$, so that the tendency to alternation is observed between 0.96 and 0.43. The pattern of Q and e changes for MCMs, permitting the conclusion that in styrene copolymerization in DMSO even the vinyl group polarity is much more significant than the mesomeric effect.

Of great interest is the emulsion copolymerization of the MCM dispersed in the form of drops sized 10^{-3} to 10^{-2} mm. As a rule, emulsions are stabilized using standard surfactant-type emulsifiers. MCM copolymerization starts in the aqueous solution and continues in emulsifier-stabilized polymer-monomer moieties. However, for water-insoluble comonomers such as styrene the topochemistry of the process is not quite clear. The emulsion copolymerization of styrene with Sr(II) acrylate is assumed to take place by the mechanism including macroradical growth in different phases.[180] In this case copolymers are mainly formed due to the recombination of these radicals at the "water-micelle" phase boundary. The product yields decrease in the order of acrylates of the following metals: Sr(II) ≈ Ca(II) < Mg(II) < Na(I) ≈ K(I). The kinetic curves for the yields of styrene—Na(I), K(I), Zn(II), and Mg(II) methacrylate copolymers in the emulsion process reach the limit at a maximum conversion of 85%.[181]. In these cases block copolymers appear, the mechanism of their formation consisting of the intramolecular interaction of polystyrene chains growing in the emulsified phase with polyacid chains growing in the aqueous phase.

Interesting features are observed in the polymerization of AN with potassium methacrylate or its complex with 18-crown-6.[182] The MCM complex represents a more electrodonor monomer as compared with the initial MCM (Q = 0.64, e = –0.85 and Q = 0.34, e = –0.54 for the complex and the potassium acrylate, respectively), the complex exhibiting a higher tendency to alternation (r_1 = 0.08, r_2 = 0.19, and $r_1 \cdot r_2$ = 0.014). These experimental data suggest the presence of numerous and not yet clearly understood tools for controlling the copolymerization parameters in MCM-based systems.

Finally, there is one more aspect in the copolymerization of ionic-type MCMs, which deserves consideration. What is meant is ternary copolymerization in which using MCMs presents interest in terms of controlling the reactivity of vinyl monomers and the spatial configuration of the final product. Thus, even small amounts (0.5 to 2.0 mol%) of alkali

Table 8 **Changes in MMA and MAA copolymerization constants on the addition of MCMs to the system**[184,185]

System	r_1	r_2	$r_1 \cdot r_2$	Harwurd parameter
MMA—MAA	0.37	0.85	0.315	36.90
MMA—MAA—$CH_2=C(CH_3)OCOLi$	0.12	0.16	0.019	66.45
MMA—MAA—$CH_2=C(CH_3)OCONa$	0.19	0.11	0.021	53.76
MMA—MAA—$CH_2=C(CH_3)OCOK$	0.27	0.08	0.022	45.05
MMA—MAA—$(CH_2=C(CH_3)OCO)_2Cu$	0.36	0.34	0.122	41.7
MMA—MAA—$(CH_2=C(CH_3)OCO)_2Co$	0.59	0	0	30.7

and alkali-earth metals added to a copolymerizable MMA-MAA system (90:10) in bulk exert a significant effect on relative reactivity constants and kinetics of the process to influence the physicochemical behavior of the appearing copolymers with the following structure:[183,184]

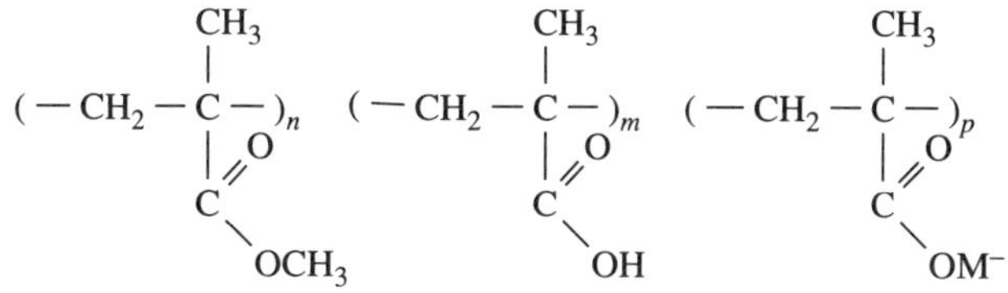

where n = 88.1 to 88.7 mol%, m = 11.12 to 11.25 mol%, and p = 0.05 to 1.94 mol%.

The addition of the MCM was found to increase the yield of the copolymerization products and their MM. These facts were attributed to MCM complexing with the copolymerizable system components, which produces a marked effect on the intermolecular interactions and promotes the formation of "blanks" with a certain monomer molecule orientation. This leads to reaction acceleration and emergence of more regular and longer macrochains. Moreover, the addition of the MCM results in a smaller $r_1 \cdot r_2$ product, which corresponds to a higher degree of monomer units alternation; the same is confirmed by a larger Harwurd parameter (Table 8).

The previous sections of this chapter were concerned with copolymerization of systems where one component is represented by a conventional monomer with the known copolymerization activity. (In the latter case there are two such components.) The most interesting kind of copolymerization is that of different types of MCMs. Among the few examples, mention should be made of copolymers with MM = 3,800 to 26,850 which are formed from sodium acrylate and sodium vinyl sulfonate,[186] as well as sodium styrene sulfonate and sodium chloracrylate in a water-alcohol environment ($r_1 = 1.39$, $r_2 = 0.26$).[187] The latter process involves a partial monomer dehylogenation to lead to intramolecular cyclization and appearance of lactone units in the copolymer. Water-soluble alternating copolymers were also obtained through the copolymerization of sodium styrene sulfonate and sodium-N-[4-sulfophenyl(maleimide)]:[188]

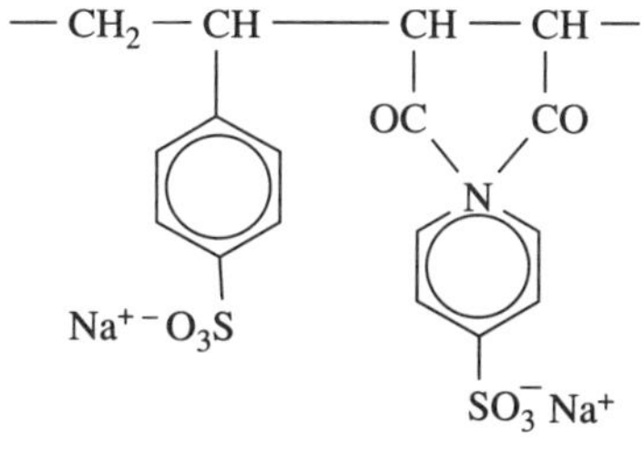

It is evident that the data available concern MCM copolymerization with one metal, namely, sodium. However, it should be expected that in the near future there will appear works devoted to the copolymerizations of MCMs containing various alkali and alkali-earth metals.

B. COPOLYMERIZATION OF NONTRANSITION METAL ACRYLATES WITH OLEFINS AND DIENES

This is a special field of ionic-type MCM copolymerization due to the use of Ziegler-Natta catalysts. Thus, fairly good results are obtained in the copolymerization of ethylene with potassium and calcium acrylates—a catalytic $(RO)_2TiCl_2$—$Al(C_2H_5)_nCl_{3-n}$ system—to afford homogeneous and well soluble copolymers.[189] In addition, the process is markedly affected by the type of radical such that MCM content increases in the series *iso*-C_3H_7 < *iso*C_4H_9 < *n*-C_4H_9, the reactivity of $KOCOCH{=}CH_2$ being twice that of $Ca(OCOCH{=}CH_2)_2$.

Copolymerizations of ethylene with acrylic acids and salts resulting from their interaction with $Al(C_2H_5)_xCl_{3-x}$ have been compared elsewhere.[73–75] The diacrylate derivative $Al(OCOCH{=}CH_2)_2Cl$ leads to the deactivation of the polymerization process, the content of the acrylate units being as low as 0.01 mol%. The best results are obtained when MCM is represented by a $CH_2{=}CHOCOAl(C_2H_5)Cl$—$Al\ (C_2H_5)Cl_2$ complex which simultaneously acts as a comonomer and a cocatalyst. In this case the content of the acrylate units is as high as 0.28 mol%.

Ethylchloraluminum acrylate is obviously incapable of copolymerization with propylene (toluene or hexane, catalyst $TiCl_3$—$Al(C_2H_5)_2Cl$, 50°C).[76] At the same time, satisfactory results are obtained using comonomers of higher unsaturated acid salts such as $[CH_2{=}CH(CH_2)_nOCO]_2AlCl$ and $CH_2{=}CH(CH_2)_n(OCOAlC_2H_5)Cl$, where n = 3 or 8.

An Al(III)-containing methacrylate $CH_2{=}C(CH)_3OCOAl$ (*iso*-$C_4H_9)_2$ (M_2) was copolymerized on a neodymium catalyst with butadiene (toluene or heptane, 25°C).[190] The relative reactivity constants r_1 = 65 and r_2 = 0.02.

C. COPOLYMERIZATION OF ORGANOTIN- AND LEAD MONOMERS

Such MCMs as Sn(IV) and PB(IV) derivatives and primarily esters of unsaturated acids undergo a fairly ready copolymerization with various vinyl monomers including styrene, AN, MMA (as well as with butadiene, cyclohexadiene, etc.) to give solid and resilient materials. The widest recognition among these MCMs used as the comonomers has been gained by trimethyl-, tributyl-, and triphenyltin methacrylates (Table 9). Thus, trimethyltin acrylate readily copolymerizes with styrene.[200] A negative value of *e* for this MCM is due to the ionic nature of the Sn—O bond and COO^- group specificity as compared with COOH or COOR. Dimethyltin dimethacrylate with styrene affords a cross-linked copolymer. Under identical conditions of TBTM copolymerization the yield of the products depended on the type of comonomers used and was equal (in percentage) to 90 (MMA), 40 (AN), 30 methacrylonitrile (MAN) and 25 (styrene).[200] Attempts to copolymerize TBTM with α-methylstyrene were unsuccessful, which was attributed to steric hindrances.

The main features of TBTM copolymerization with alkyl methacrylates reside in the fact that these comonomers are distributed in the chain randomly, the alternation tendency increasing with alkyl chain length in the case of both acrylates and methacrylates.[204] The same is true of the copolymerization of other tributyltinalkyl acrylates. Due to a crossed growth, the rate of this reaction is higher than that in the case of "metal-free" analogues as a result of a crossed growth. Moreover, an increase in the bulk of the side groups and formation of coordination bonds between the monomer units appears to handicap macroradical interaction and to create favorable conditions for chain propagation. A

Table 9 **Copolymerization parameters of Sn(IV)-containing ionic-type monomers**

MCM (M_2)/Q-e parameters	Parameter	Comonomer: Styrene	MMA	VA	MA	BA	BMA	AN	VPr	MMA	Allyl methacrylate	Glycydyl methacrylate	Ref.
$(C_4H_9)_3SnOCOCH{=}CH_2$	r_1	0.21	2.2	0.06	0.82	0.70	1.65	1.01	0.610	–	0.20	4.29	191–195
			1.0										
$Q = 0.235$, $e = 0.401$	r_2	1.91	0.4	2.57	0.03	0.15	0.20	0.24	0.513	–	2.26	0.34	
$(CH_3)_3SnOCOC(CH_3){=}CH_2$	r_1	1.57	–	–	–	–	–	–	–	0	–	–	8, 196
$Q = 0.45$, $e = -0.37$	r_2	0.53	–	–	–	–	–	–	–	0.22	–	–	
$Q = 0.31$, $e = -0.64$													
$(CH_3)_2Sn[OCOC(CH_3){=}CH_2]_2$	r_1	0.28	–	–	–	–	–	–	–	–	–	–	196
$Q = 1.36$, $e = 0.41$	r_2	0.83	–	–	–	–	–	–	–	–	–	–	
$(C_4H_9)_3SnOCOC(CH_3){=}CH_2$	r_1	1.14	1.04	0.017	0.66	0.57	0.68	0.47	0.44	0	1.01	0.79	8, 192
		0.51	1.0	0.028			1.65						
$Q = 0.852$, $e = 0.197$		1.10		0.08									194
$Q = 0.18$, $e = -0.88$													
$Q = 0.78$, $e = 0.38$	r_2	0.40	0.93	4.408	1.75	0.85	0.65	0.47	3.16	0.05	2.30	0.75	197–199
$Q = 0.15$, $e = 1.51$		0.49	0.79	0.76			0.20						
		0.26		0.61									
$(C_6H_5)_3SnOCOC(CH_3){=}CH_2$	r_1	0.47	1.09	–	0.63	0.35	0.78	0.16	0.36	–	–	–	200, 201
	r_2	0.76	1.08	–	2.48	1.17	0.64	0.69	1.22	–	–	–	
$(C_4H_9)_3SnOCOC({=}CH_2)CH_2COOSn(C_4H_9)_3$	r_1	0.68	1.64	–	0.67	0.96	–	0.43	–	–	–	–	116, 117
	r_2	0.23	0.22	–	0.39	0.52	–	0.44	–	–	–	–	
$CH_2{=}CHC({=}O)OC_6H_4COOSn(C_4H_9)_3$	r_1	1.34	1.72	–	1.05	2.11	2.83	–	–	–	–	–	81

Table 9 (continued)

MCM (M_2)/*Q-e* parameters	Para-meter	Comonomer											Ref.
		Sty-rene	MMA	VA	MA	BA	BMA	AN	VPr	MMA	Allyl metha-crylate	Glycy-dyl metha-crylate	
$Q = 0.045$, $e = 1.39$	r_2	0.12	0.16	–	0.10	0.02	0.01	–	–	–	–	–	
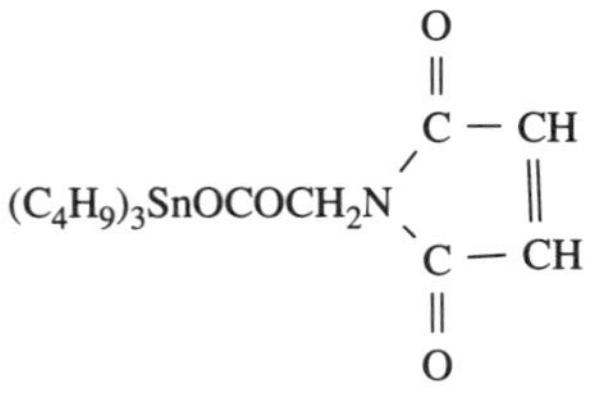	r_1 r_2	1.65 0.004	– –	– –	– –	– –	– –	– –	– –	0.02 0.16	– –	– –	115
$(C_4H_9)_3SnOCOCH{=}CHC(=O)OCH_2CH{=}CH_2$	r_1	0.12	–	–	–	–	–	–	–	–	–	–	202
$Q = 1.11$, $e = 1.67$	r_2	0.018	–	–	–	–	–	–	–	–	–	–	
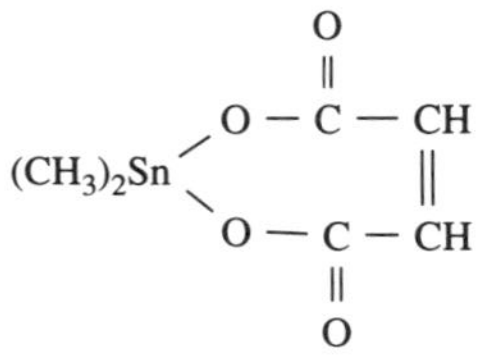	r_1 r_2	– –	– –	– –	2.08 0.22	– –	– –	– –	– –	– –	– –	– –	114
$Q = 0.24$, $e = 0.08$													

larger Q value of tin-containing monomers as compared with that of the corresponding alkyl methacrylates points to the impact of the electron-acceptor SnR_3 groups exhibiting a strong induction effect on the double bond conjugation with the carbonyl group. The alternation of the units—supported by r_1 and r_2 values as well as by the S-shaped curves of polymer compositions and other factors—is caused by a negative double bond polarization as a result of a fairly strong negative charge localization at the oxygen, the charge being inductively transferred toward the double bond:[191]

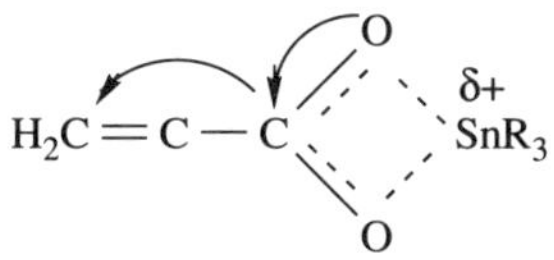

This pattern of charge density distribution agrees with an assumption about a high Sn—O bond polarity.

Elongation of the tin-associated alkyl chain enhances the tendency to intermolecular coordination and monomer alternation, e.g., in copolymerizations with MMA.[8,191] Accordingly, the Q and e values become smaller; this can be attributed to the formation of intermolecular complexes from the parent monomers with the result that —irrespective of monomer mixture composition—the structure of the resulting copolymer—in terms of monomeric units—is close to 1:1.

It should be noted that the contribution of the complexing process to the Sn(IV)-based MCM copolymerization is more pronounced than in the case of homopolymerization. This problem has received much attention, particularly, in the copolymerization of TBTM with MMA,[205] allyl chloracetate,[206] tributyltinallyl maleate with styrene,[202] tributyltin acetate maleimide with styrene and maleic anhydride,[115] etc. In copolymerizable MMA-TBTM or styrene-TBTM systems one can observe the occurrence of complex donor-acceptor interactions involving double bond π-electrons, the decisive role being played by a pentacoordinated state of tin. In the transition state the electrons are more or less readily transferred from TBTM to MMA or from styrene to tributyltinallyl maleate:

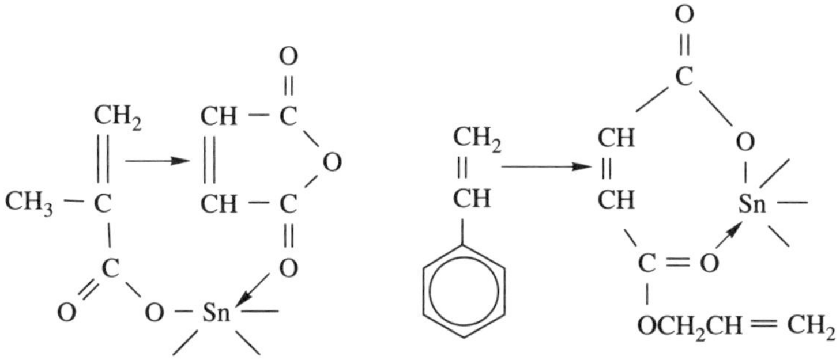

For the first complex, the complexing constant K = 0.17 l/mol.[205] The formation of such a complex has been demonstrated by calorimetry. Also, formation of complexes in a tributyltin acetate maleimide-maleic anhydride system with the equilibrium constant equal to 0.005 l/mol has been reported.[115]

In the case under discussion chain propagation can take place through three mechanisms, namely, a consecutive addition of free monomers to the macroradical, addition of complex-bonded monomer pairs, and simultaneous participation of the free and complex-bonded monomers. Styrene-tributyltinallyl maleate copolymerization has been used to exemplify the case when reactivity of the complex-bonded monomers is higher than that of the free ones. Their predominant contribution is observed in reactions with propagating radicals irrespective of the type of terminal unit; however, the addition of free molecules to the propagating macroradical is probable.[202] It is interesting that in the case of copolymerization with an allyl-type monomer, i.e., allyl chloracetate, tin coordination with carbonyl acetate leads to a significant suppression of degradative chain transfer with the result that allyl chloracetate is involved in the chain propagation process in the intermediate complex:[206]

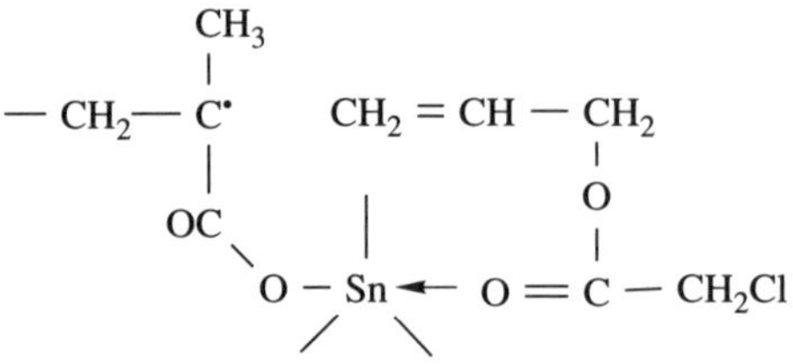

In copolymerization with a large number of conventional monomers, TBTM acts as a more reactive monomer, most pairs exhibiting a tendency to alternation (see Tables 2–9). Such a tendency is also displayed by other Sn(IV)-containing monomers. Thus, in the copolymerization of *p*-acryloxytributyltin benzoate with various vinyl comonomers, the resulting products always incorporated large blocks of comonomer units separated by MCM units.[191] The possibility of controlling the composition of copolymers based on tin-containing monomers can be realized by changing conditions of the process. Thus, when MMA is copolymerized with TBTM in emulsion rather than in solution, the product is enriched with MMA units at any monomer mixture composition.[191] This is associated with the fact that in water the solubility of MMA is higher than that of the MCM. In water- and emulsion-occurring copolymerizations the r_1, r_2, Q_2 and e_2 values are 0.82, 0.03, 0.25, –0.5; and 1.97, 0.51, 0.23, –0.1, respectively.

The way tercopolymers based on Sn(IV)-containing monomers are formed depends on the copolymerization constants for the binary systems of various monomer pairs. This is exemplified by the following systems: TBTM—MMA—AN,[193] TBTA—VA—VPr, TBTA—VA—AN, TBTM—VA—AN, and TBTM—MMA—VPr,[194] triphenyltin methacrylate-styrene,[200] and di(tributyltin)itaconate-AN with methyl-, ethyl-, butyl acrylates, and with styrene.[116]

Among the other ionic-type Sn(IV)-containing MCMs that represent components of copolymerizable systems, mentioned should be made of successful copolymerizations of dibutylchlorotin acrylate (maleate, cinnamate) with methylacrylate (MA) and vinyl chloride,[207] as well as nonhomopolymerizable tributyltin maleinate and fumarate with styrene and MMA.[114,208] Regarding lead-containing MCMs of this type, there are reports on the copolymerization of di- and trialkyl(phenyl)plumbic acrylates[24] and methacrylates.[209]

D. COPOLYMERIZATION OF IONIC-TYPE MCMs BASED ON TRANSITION METALS

The rate of radical styrene copolymerization with transition metal acrylates (in methanol, AIBN initiation) increases with MCM fraction in the monomer mixture.[152,210,211] However, Cu(II) acrylate produces a practically inhibiting effect on the process even at its low contents. Table 10 indicates that the copolymerization constant of styrene is nearly always higher than that of MCMs with the tendency of a random distribution of units in the copolymer chain.

The negative values of *e* (see Table 10) point to an electron-donor nature of the metallocarboxyl group in the MCM, the drop of this value being correlated with increasing metal electronegativity (note for comparison that in the case of styrene-AA copolymerization $e_2 = 0.77$). In addition, a lower MCM reactivity is corroborated by the fact that $Q_1 < 1$, which is also true of the $1/r_1$ quantity affecting acrylate reactivity with respect to the polystyrene radical.

Table 10 **Copolymerization parameters of styrene and AN with transition metal acrylates**[210–214]

M_1	Acrylate (M_2)	r_1	r_2	$r_1 \cdot r_2$	$1/r_1$	Q_2	e_2
Styrene	Co(II)	1.74	0.56	0.97	0.58	0.51	–0.64
Styrene	Ni(II)	1.83	0.53	0.97	0.55	0.48	–0.63
Styrene	Cu(II)	5.94	0.12	0.71	0.17	0.11	–0.22
Styrene	Zn(II)	1.10	0.90	0.99	0.91	0.84	–0.70
AN	Co(II)	0.14	0.16	0.022	7.1	0.42	–0.75
AN	Ni(II)	0.10	0.17	0.017	10	0.59	–0.85
AN	Cu(II)	0.21	0.08	0.017	4.8	0.26	–0.82
AN	Zn(II)	0.41	0.24	0.106	2.44	0.24	–0.30

Note: Copolymerization conditions:with styrene in methanol, with AN in DMPA.

Such a behavior was also observed in the copolymerization of these MCMs with AN in DMFA at 60°C (Table 10).[212–214] It is evident that there is a clear tendency toward alternation. The process occurred in solution until the MCM content in the monomer mixture became 12 mol%; at AN concentration under 30 mol% the copolymer partially precipitated.

Generally, the copolymerization (as well as homopolymerization) of diacrylates can involve some unsaturated bonds being retained in the product; this is the case, for example, in copolymerization with styrene:

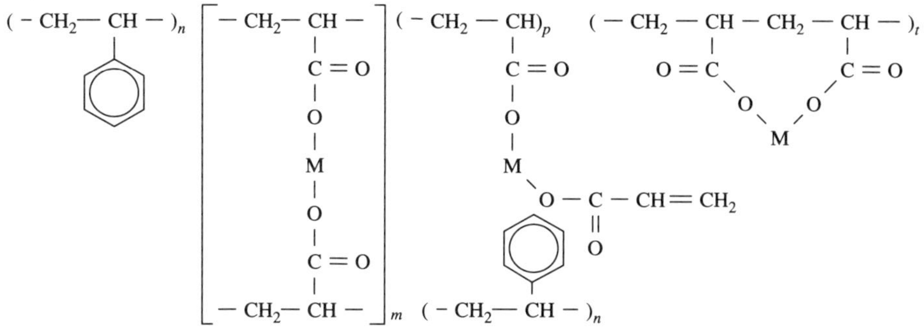

For such divinyl MCMs the copolymerization can occur by two different mechanisms. According to the first mechanism, the propagating chain interacts with one of the acrylate residues only, followed by comonomer addition. At a certain stage the process includes the second acrylate residue to give rise to a cross-linked chain:

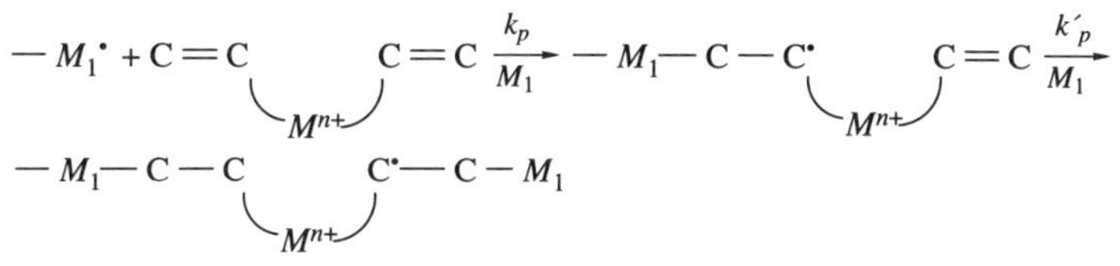

If the activity of the second residue increases in the presence of the propagating radical ($k_p' > k_p$) in the same complex, the second mechanism (i.e., its intramolecular addition to the resulting radical and formation of a linear-type polymer) will apply:

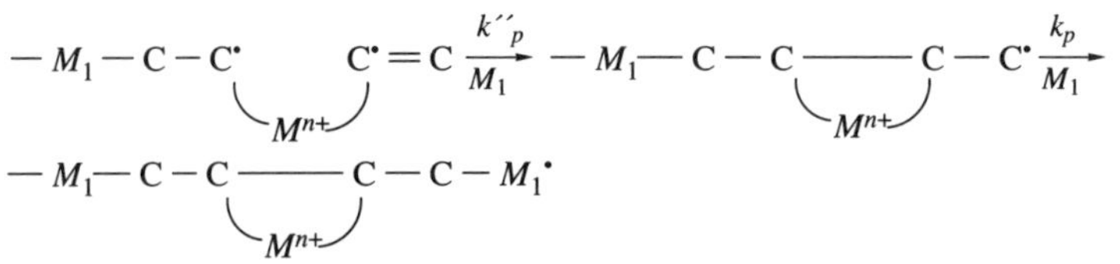

The contribution of each of these mechanisms appears to depend on a number of factors and can only be assessed after studying the structures of the resulting polymers. Special studies indicate that the products appearing at the initial copolymerization stages are soluble in organic solvents.[210] Consequently in this case, no intermolecular cross-linking due to the vinyl groups takes place, and the larger portion of these groups interacts with the two acrylate groups. However, at higher conversions unsaturation becomes clearly pronounced. Thus, the number of unreacted double bonds in the copolymers grows in the acrylates of these metals as follows: Zn(II) (4 to 35%) < Co(II) (14 to 39%) < Ni(II) (22 to 49%) (unsaturation is 50% if no more than one acrylate group reacts). This series correlates with the reactivity of the double bond of the acrylates in their homopolymerizations. As will be shown below (see Chapter 6), the residual unsaturated groups can be involved in further polymerization conversions.

The distribution of units in the macromolecular chain constitutes an important characteristic of the copolymer, which defines their properties. The r_1 and r_2 values were useful in calculating the probabilities of structure formation (—M_1—M_1—, —M_1—M_2—, —M_2—M_2—) and the corresponding functions of the copolymerization chain units distribution.[215] It is evident from Table 11 that in the Ni(II) acrylate-styrene copolymer the regularly alternating structures are responsible for 46% of the entire number of all the acrylate units.

In the preceding sections consideration was given to the copolymerization of transition metal acrylates in nonaqueous solutions with practically no MCM dissociation. A more complicated situation exists in the case of a process taking place in aqueous environments, which has been exemplified by a methacrylatopentaamminocobalt(III) perchlorate-MAA or *cis*-dimethacrylatotetraamminocobalt(III) perchlorate-MAA system.[88] The composition curves at pH of the reaction environment equal to 3.0 and 7.0 are markedly different in that the MCM gets inserted into the copolymer chain more readily at pH 7.0. As in the case of homopolymerization, this is due to the electrostatic interaction of the growing radical and monomer: at pH 7.0 the major part of the MCM is dissociated. Therefore, the addition of the methacrylate anion to the likely charged radical will be handicapped. At the same time, there arise conditions for the addition of the MCM dissociated at the perchlorate anion.

For the dissociated and undissociated MAA, copolymerization parameters differ markedly. The Q_2 and e_2 values are significantly lower when the process runs at pH 3. The fact that the MCM fraction in the copolymer may exceed 50 mol% implies that among other alternatives there may exist sequences containing at least two MCM units, despite the inherent (and unavoidable) steric hindrances.

It has been noted with a particular reference to alkaline metal methacrylates that when present even in small amounts, they produce a noticeable effect on MMA and MAA block copolymerization. It is still more so for the Cu(II) and Co(II) acrylates.[85, 216] The addition of 0.15 to 1.5 mol% of a Cu(II) salt to the reaction mixture leads to an appreciable increase in the reaction rate with the appearance of an induction period showing a growth as the amount of the additive is increased (Figure 11). In contrast to this, the Co(II) acrylate accelerates the reaction. In its presence the r_1 and r_2 values were found to show a marked change (Table 8) with a parallel growth of tendency toward alternation. At the same time, a remarkably suprising fact is associated with a high Cu(II) activity in the case of block copolymerization with MMA with no MAA additives (BP, 60°C).[217] Furthermore, a

Table 11 **Distribution of monomeric units in the series of styrene copolymers (M_1) with metal acrylates**[215]

Unit	Content of metal acrylate (mole fraction)	
	Ni(II)	Co(II)
—M_1—	0.25	0.11
—M_1M_1—	0.10	0.13
—$M_1M_1M_1$—	0.03	0.11
Remaining M_1 units	0.02	0.06
Total number of M_1 units	0.41	0.41
—M_2—	0.25	0.17
—M_2M_2—	0.15	0.11
—$M_2M_2M_2$—	0.06	0.06
Remaining M_2 units	0.13	0.07
Total number of M_2 units	0.59	0.59

correlation with the data on the Cr(II) copolymerization in similar conditions displays a fairly close activity of these MCMs (M_2) : $r_1 = 1.05$, $r_2 = 0.8$; $r_1 = 1.08$, $r_2 = 0.95$ for the Cu(II) and Cr(III) acrylates, respectively.

Comparatively less popular is an efficient method such as the addition of components to obtain MCMs *in situ*. In this context, we shall mention the production of an iron-containing PAA,[219] distinguished by a "relay-type" formation of a MCM in the presence of the Mohr salt, a component of the redox system, and its copolymerization with AA at 25 to 40°C:

$$\underset{\displaystyle \text{COOH}}{\text{CH}_2{=}\underset{|}{\text{CH}}} \xrightarrow[\text{H}_2\text{O}]{\text{K}_2\text{S}_2\text{O}_8,\ \text{FeSo}_4,\ (\text{NH}_4)_2\text{SO}_4} (-\text{CH}_2-\underset{\displaystyle \text{COOH}}{\underset{|}{\text{CH}}}-)_m\ (-\text{CH}_2-\underset{\displaystyle \text{COOFe}_{1/n}}{\underset{|}{\text{CH}}}-)$$

where $n = 2$ or 3 and $m > 150$. The MM of such a product is $5.5 \cdot 10^5$ to $2 \cdot 10^6$. A similar method was used to obtain a Fe(III)-containing polymethacrylic acid (PMAA) from MAA and $FeCl_3$ followed by irradiation of the product in the solid phase.[220] The metal was responsible for 1 wt.% in the copolymer.

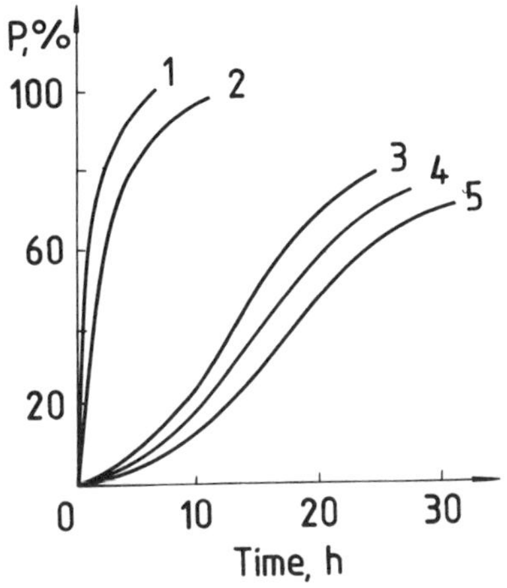

Figure 11 Effect of Cu(II) methacrylate additive on polymerization in MMA–MAA system. [MCM] wt%: 0.1 (1), 0.3 (2), 0.5 (3), 1.0 (4), and 1.5 (5).

Note also the available data on the copolymerization of tributoxytitanium methacrylate with MMA (BP, 90°C), as well as dibutoxytitanium dimethacrylate and -maleate with styrene.[90] Copolymerization of methacryloxytitanocene chloride and dimethacryloxytitanocene with styrene and MMA corroborates a high activity of such MCMs in the transfer and termination of the growing chains.[96] The activity of Ti(IV)-containing macromers such as polybutoxytitanoxafuryl acrylate and polybutoxytitanium cinnamate were copolymerizable (with styrene): at 100°C they yield transparent red-brown copolymers.[91–93] There are also reports on the copolymerizations of zirconyl acrylate and methacrylate with styrene and a variety of alkyl acrylates with the appearance of cross-linked copolymers.[221]

Copolymerizations of transition metal acrylates were exemplified at great length above. However, other MCM types can be used for the same purpose. For instance, there are data on the effective copolymerization of ethylene glycol methacrylate phthalate salts with the general formula:

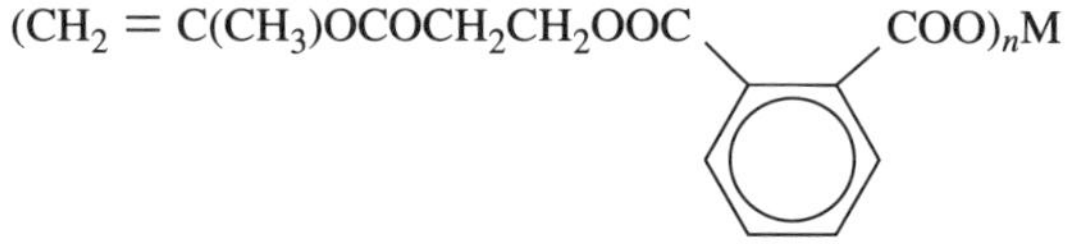

where M stands for both transition[124] and nontransition metals,[123] with styrene or MMA in the presence of a redox methyl ethyl ketone-cobalt naphthenate system. Most interesting are copolymers based on a Cr(III)-containing monomer. The process can best be conducted in two stages: stage 1—at 55°C for 10 h and stage 2—at 100°C for 5 h. In all cases a system was found to gelatinize at a 10 to 20% conversion.

E. COPOLYMERIZATION OF IONIC-TYPE MCMs BASED ON LANTHANIDES AND ACTINIDES

This field of MCM polymerization conversions is still at the embryonic stage; however, there is every reason to expect considerable progress in this area in the near future. There are reports on the production of tercopolymers composed of Eu(III) as well as Tb(III) methacrylates and of MAA and MMA.[98] An MCM was dissolved in MAA and then added to a preoligomerized MMA (AIBN initiation). The mixture was compacted and heated at 45 to 105°C in inert atmosphere. The resulting tercopolymers comprised clear plates.

Virtually the same method—and also without preoligomerization with the use of a large number of comonomers such as MMA, styrene, 2-hydroxyethyl acrylate, and Pb(II) methacrylate in various combinations—has been instrumental in producing binary, ternary, and quarternary Nd(III)methacrylate copolymers with a metal content from 0.1 to 5 wt%.[97] Ternary copolymers of Eu(III) and Tb(III) triundecylenates (AIBN initiation, in block) were obtained in the form of clear plates.

Bismethacrylate uranyl dichloride readily copolymerizes with MMA and MAA in the presence of AIBN.[100] This MCM was shown to initiate MMA polymerization in the presence of *tert*-butyl hydroperoxide even at room temperature. In this case the process is inhibited by air oxygen.

F. GRAFT POLYMERIZATION OF IONIC-TYPE MCMs

Ionic-type MCMs can be graft-polymerized using a large variety of approaches. Thus, there are standard redox systems such as Ce(IV)–HNO_3, Fe(II)–H_2O_2, and $(NH_4)_2S_2O_8$–$Na_2S_2O_3$ leading to the formation of an immobilized radical due to the abstraction of one of the substrate hydrogen atoms:

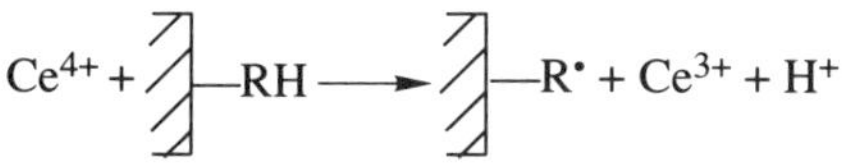

However, in this case a radical is likely to appear not only on the substrate but also in the solution as well so that the products usually contain a significant amount of homopolymer.

An $(NH_4)_2S_2O_8$–$Na_2S_2O_3$ initiating system was useful in grafting sodium acrylate on a polysaccharide (xylan).[222] On removal of the homopolymer, a graft copolymer containing 96.5% of sodium polyacrylate and 3.5% of xylan was obtained. A more complex initiating system, isopropylbenzene-hydroquinone-sodium sulfide hydroperodixe, was used for grafting sodium methacrylate on acetylcellulose.[223] In the conditions under study (an aqueous-acetone solution, 15 to 50°C), the process resulted in nearly 100% conversion of the polymer grafted and a high overall efficiency of the process (90 to 95%) generally not typical of chemical initiation. It is interesting that in this case, insoluble spherical particles of the grafted polymer appear in the reaction mixture, which get isolated into a separate disperse phase.

A somewhat different approach was used for grafting sodium acrylate and methacrylate on cellulose. In this case a cotton fabric was impregnated with an aqueous $(NH_4)_2S_2O_8$ solution, treated with a solution of this MCM and heated at 105 to 120°C.[224] In this case, monomer conversion was found to linearly increase with time. In the case of sodium methacrylate, a temperature increase led to higher polymer yields, whereas with the acrylate, the effect was just the reverse.

A redox Fe(II)-H_2O_2 system was also used to effect graft polymerization of Ca(II) acrylate on cellulose.[225] When use was made of a 20% MCM solution, the reaction ran at a high rate; and at 20°C it came to an end in 5 min with no yield growth as the temperature was increased. Note that decreasing MCM concentrations to 16% resulted in a sharp fall of conversion.

Graft polymerization can take place not only in solutions. Thus, it has been proposed that Zn(II) acrylate and other AA and MAA salts be grafted on polyolefins in a melted state at 100 to 120°C in a mixer in the presence of a peroxide initiator.[226] However, in this case the homopolymer content in the product was not determined.

Chemical initiation also proved effective in MCM graft polymerization. Thus, a mixture of sodium 2-acrylamido-2-methyl-propane sulfonate and AAm were grafted on dextran in the presence of a Ce(IV)-HNO_3 system.[227] In this case the grafting process was complicated by Ce(IV) being bonded to the MCM anion; therefore, the best results were attained after Ce(IV) had been prefixed on the dextran. The conversion of the MCM did not exceed 80%; whereas for AAm, it was as high as 100%. The M_w of the grafted component was about $2 \cdot 10^6$. Remarkably, the content of the grafted polymer differed from that of the product resulting from the radical copolymerization of AAm and this MCM in solution.

Homopolymer content in the product is also enhanced by radiation initiation of graft polymerization through a direct exposure of the reaction system. However, the method of running the process is rather simple and therefore was applied to MCM grafting. Thus, plasmochemical initiation was found to be instrumental in sodium acrylate and vinyl sulfonate grafting on wool fiber and fabric.[228] The MCM was applied on the substrate from an aqueous solution to the time its weight increased by 10 to 20% of the initial one. On irradiation and extraction of the unreacted MCM, the degree of grafting was found to equal 2 to 5%. To modify PVC fibers, a radiation grafting technique has been employed.[229] The fibers were immersed in a mixture of a Ca(II) acrylate solution and H_2O—CH_3OH, containing the appropriate inhibitors such as Fe(II), Fe(III), Cu(I), and Cu(II) salts. To the solution was added $C_2H_2Cl_2$ until separation of the organic and aqueous phases; and then the fiber was exposed. This gave graft copolymers in which the Ca(II) content exceeded 15%. Another MCM, triethyltin methacrylate, was grafted on various substrates (such as polyethylene [PE], polypropylene [PP], polyamide, and poly(ethylene terephthalate) [PETP]; and wooden materials in the form of films, tubes, and bars) by immersion of a deaerated MCM solution into an organic solvent (acetone, benzene, or methanol), followed by exposure to 60°C radiation (I = 0.003 to 0.50 J/kg·s) to a dose of 1 to 150 kJ/kg.[230] The degree of grafting was 0.5 to 35.2 wt %.

More promises are offered by postradiation methods of graft polymerization when the stages of radiation and grafting proper are time-separated. In this case, the probability of homopolymer formation is much lower. Thus, good results are attained by using this method for grafting sodium vinyl- and *p*-styryl sulfonates on PE films (electron accelerator irradiation).[231] However, attempts to effect grafting on unmodified substrates were unsuccessful; and the process was only effective when the polymer was prehydrophilated, e.g., through AA grafting.

Far more useful is graft polymerization by preirradiation in air. At 70 to 100°C the resulting hydroperoxide (HP) groups decompose to give alkoxy radicals which are capable of initiating the graft process:

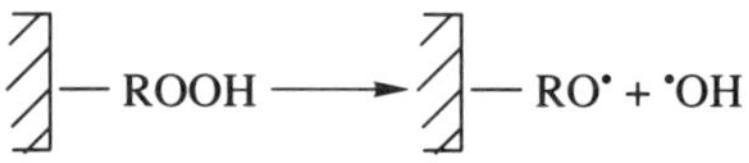

The degree of the resulting grafting of triethyltin methacrylate on PP (D = 10 kJ/kg, I = 0.010 J/kg·s, acetone, 60°C) was as high as 9 wt %.[230] Acrylates of transition metals such as Cr(III), Fe(III), Co(II), Ni(II), and Cu(II) were grafted by boiling a suspension of a powdered PP in alcohol.[232,233] The grafting of such MCMs is distinguished by the practical absence of homopolymer in the product. At the same time in the case of potassium acrylate graft polymerization, the homopolymer content in the product was as high as 30%. It is obvious that ions of varying valence metals strongly impede the stage of homopolymerization initiation. For all the acrylates under study one can observe similar kinetic dependences of the process in that during the first 30 min the rate is maximum and then it drops sharply (Figure 12). A twofold increase in the preirradiation dose makes the

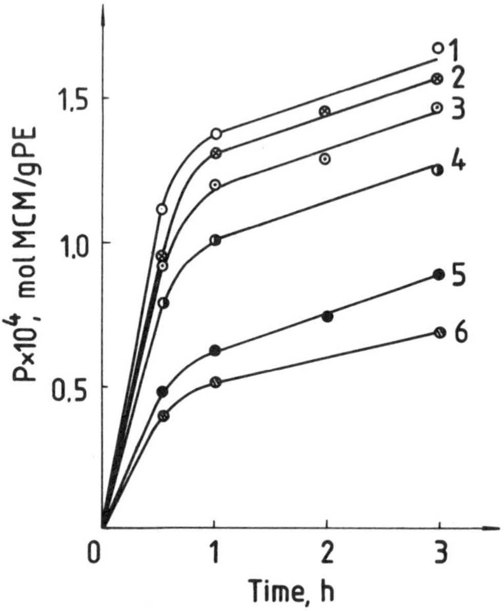

Figure 12 Kinetic curves for yield of graft metal acrylate polymers: Co(II) (1, 3), Cu(II) (2), Ni(II) (4), Cr(III) (5), and Fe(III) (6). Ethanol, [MCM] = 0.4 mol/l; 79°C; D, kJ/kg: 200 (1, 2, 5, 6), 100 (2, 3, 4); I = 1.4 J/kg·s (2, 5, 6), I = 2.8 J/kg·s (1, 3, 4).

initial grafting rate grow by no more than 15 to 20% (curves 1 and 3). Note that the grafting rate of the divalent metal acrylates is higher than that of the trivalent ones. As this takes place, the activities change as follows: Co(II) > Cu(II) > Ni(II) > Cr(III) > Fe(III). It is evident that the relation between the Co(II) and Ni(II) acrylate activities is similar to that observed in the case of radical homopolymerization,[83] as well as in the copolymerization of these MCMs using conventional monomers.[210,211] It is interesting that the grafting of Cu(II) acrylate occurs rather effectively as distinct from the polymerization of this MCM in solution.

In the concentration range 0.04 to 0.40 mol/l the initial rate of graft polymerization for all the acrylates under study is essentially constant. This is associated with the limiting MCM concentration being attained on the PE powder surface. This concentration corresponds to monolayer coverage.

The acrylate grafting kinetics is substantially affected by the nature of transition metals. A comparison of the initial grafting rates and HP concentration in the irradiated PE suggests that their interrelation can be presented as $w_0 \sim [HP]^{\alpha}$ at $\alpha = 0.7, 0.6, 0.5, 1.1$, and 1.1 for grafting the Co(II), Ni(II), Cr(III), Cu(II), and Fe(III) acrylates, respectively. The α-values obtained point to a predominantly bimolecular termination of the growing polymer chains in the polymerization of the Co(II), Ni(II), and Cr(III)-containing MCMs and a monomolecular mechanism in the case of the Cu(II) and Fe(III)-containing MCMs. A specific behavior of the latter was discussed above.

Metal acrylate grafting is distinguished by moderate E_a values, from 42 to 55 kJ/mol, as well as by a weak dependence of the process rate on the presence of radical reaction inhibitors in the system. The latter seem to complex with the MCMs and therefore display no activity in the inhibition processes.

G. MUTUAL MCM COPOLYMERIZATION

The mutual copolymerization of various MCMs aimed at producing polymers containing different metals, is one of the most promising areas in the solution to the problem under discussion. To begin with, no systematic studies of this subject have been made yet. In the preceding sections we made mention of the copolymerization of ionic-type sodium-containing MCMs as well as Nd(III) and Pb(II) methacrylates.[97] In this context, we shall pay special attention to the copolymerization of methacrylatopentaamminocobalt(III) perchlorate with sodium styryl sulfonate (pH 1.0, 50°C, $K_2S_2O_8$).[238] This process yields a polyampholyte in which the sodium salt content (60 to 70%) is independent of reaction mixture composition. Thus, the resulting copolymer is characterized by the presence of a repeating unit emerging from the polymerization of the intermediate complex:

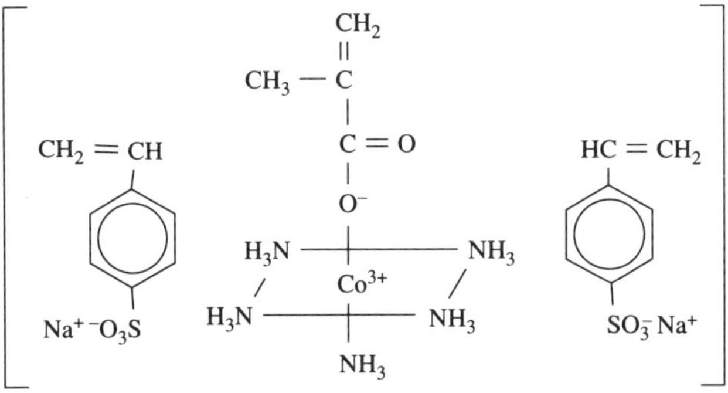

The reasons responsible for the formation of this complex are of the electrostatic nature and are similar to those that were discussed above for the copolymerization of ionizable monomers. Note also the appearance of a polymer containing, along with unsubstituted PMAA, Zn(II) and Fe(III) polymethacrylate units.[220] A long-range interaction between

Table 12 **Parameters for copolymerization of transition metal acrylates (M_1) with $(C_5H_5)_2Ti(OCOC(CH_3){=}CH_2)_2$**[215]

M_1	r_1	r_2	Q_1	Q_2	e_1	e_2
Ni(II) acrylate	0.95	0.56	0.53	0.56	–2.63	–1.9
Cu(II) acrylate	1.09	0.89	0.05	0.56	–2.13	–1.9

Zn(II) and Fe(III) ions in the copolymer, transferred over the polymeric chain, is assumed to take place.

The mutual polymerization of transition metal-containing MCMs has been studied in more detail. Thus, radical copolymerization of dicyclopentadienyltitanium dimethacrylate with Ni(II) and Cu(II) acrylates takes place relatively fast in both the liquid (DMPA, AIBN) and solid (high-pressure initiation in combination with shear strains) phases.[168,215] Although these systems are distinguished by close relative reactivity constants, the *e* values do not correspond to those of the conventional monomers in that metal-containing groups behave as strong electrodonor substituents (*e* lie between –1.96 and –2.65). The MCM units are distributed over the chain randomly. Copolymerization parameters of these MCMs are listed in Table 12. The absence of any significant differences between the copolymerization parameters of the processes in the liquid phase and plastic flow conditions may suggest that mechanisms of the two processes are fairly similar.

The method consisting of the application of high-pressure and shear strains has been used to effect tercopolymerization of Y(III), Ba(III), and Cu(II) acrylates in a 1:2:3 molar ratio.[235] The process conducted in these conditions was the only tool suitable for inserting a Cu(II)-containing MCM which is usually lowly active in polymerization conversions into the copolymer.

To sum it up, ionic-type MCMs represent a fairly well studied class of monomers. Considerable advances have been made in both development of methods for the syntheses of such MCMs and drawing them in homo- and copolymerization, as well as graft polymerization. Remarkable progress has been made in acrylate studies and more modest results, for maleic and itaconic acids. As a matter of fact, the existing methods enable one to obtain ionic-type MCMs based on almost any metal of the Periodic Table. Such monomers, except for perhaps Ti(IV) compounds, are stable in air; and even in the presence of water all of them can be stored for a long time (months or more) without any subsequent changes.

These methods, both conventional and recent, may be very convenient for the production of metal-containing polymers with a large variety of MM and, as will be shown in Chapter 5, with a specified stereostructure. At the same time, we must mention those complications that are often inherent in such processes, such as MCM dissociation and redox reactions which involve metal ions. Many of the side reactions can be avoided by using nonconventional methods of polymerization conversions.

Copolymerizations of the MCMs under discussion with conventional comonomers enlarge the possibilities of inserting metal ions into the macrochain. For some MCM, such as tin acrylates and unsaturated acid salts containing alkali and alkaline-earth metals, copolymerization studies have progressed to the stage of determining relative reactivity constants (the parameters *Q* and *e*). For other MCMs the work done has been mostly nonsystematic. It should be pointed out that much work is yet to be done in the copolymerization of transition metal-containing MCMs. These are the processes that harbor numerous, and not yet completely comprehended, tools for controlling the processes involved.

Chapter 3

Donor-Acceptor-Type MCMs

Formation of such metal-containing monomers (MCMs) (*nV*-type complexes) is due to the filling of the low-energy vacant orbitals of metals (*V*-acids in Mullicken's nomenclature)[236] with *n*-lone electron pairs of oxygen, nitrogen, phosphorus, and other heteroatoms in such groups as amine nitrile, amide, ester, etc. Monomers based on Groups IVA-VIIA and VIII of the Periodic Table represent the largest class of donor-acceptor MCMs.

I. SYNTHESIS OF DONOR-ACCEPTOR-TYPE MCMs

In principle, the MCMs of this class are conventional monomers of metal compounds with organic ligands, distinguished by the presence of a reactive multiple bond(sometimes a cycle) in a given ligand. Usually, their synthesis is far simpler than that of the previously discussed monomers. Frequently, it resides merely in mixing the calculated amounts of a complexing agent and a ligand in a suitable solvent. The optimal solvent is at the same time a precipitator for the resulting complex. If, on the other hand, there is no precipitation, the solution can be partially evaporated and then cooled or precipitated into another solvent. More complicated techniques have been employed to obtain MCMs based on platinum group metals.

Let us describe consistently the methods used to synthesize MCMs incorporating nitrogen, oxygen, phosphorus, and sulfur heteroatoms.

A. VINYLPYRIDINE-CONTAINING MCMs

Various vinylpyridines such as 4-vinylpyridine (4-VP), 2-vinylpyridine (2-VP), and 2-methyl-5-vinylpyridine (MVP) are characterized by the presence of a lone electron pair at the nitrogen atom with distinct basic properties. Coordination at this atom has been proved conclusively by infrared (IR) spectroscopy to show that in some complexes the absorption band of the C—N bond is shifted to 20/cm, whereas $\nu_{C=C}$ remains practically unaltered. Thus, for Zn(II) complexes with a variety of isomeric vinylpyridines, the $\nu_{C=C}$ shifts did not exceed 1 to 6/cm.[237]

In the majority of cases vinylpyridine complexes have the composition of metal compound $(MX_n)\cdot 2$ ligand(L).[238, 239] Thus, two modifications of the $Zn(4\text{-}VP)_2Cl_2$ complex have been obtained.[238] In the two species (α and β) the coordination with respect to the metal ion is tetrahedral with perfectly different lattice parameters. A mixture of these species was obtained by recrystallization. However, when heated to over 80°C, the α-species undergoes an irreversible solid-phase conversion to the β-species.

In the case of vinylpyridine (VP) complexes with transition metal salts, one can observe a broader spectrum of both compositions and space structures. Thus, Ti(IV), V(III), V(IV), and V(V) halides with 2-VP give adducts $MX_n\cdot 2L$ and $MX_n\cdot 3L$.[240] The synthesis of these MCMs is conducted in inert atmosphere by mixing dilute heptane solutions of the corresponding components (in the case of VCl_4 in a CCl_4 solution). For the Co(II) and Ni(II) salts, the $M(VP)_4X_2$ and $M(VP)_4X_2$ complexes are typical. The former can have both a tetrahedral (blue) and an octahedral (pink) configuration; whereas the latter has an octahedral configuration only.[238] On repeated recrystallizations from ethanol the synthesized tetrahedral α-modification converts to the β-octahedral species which is completed with chlorine bridges. The α- and β-modifications have different distances between the C=C bonds, equal to 4.47 and 3.65 Å, respectively. Note a convenience of a reverse conversion, when the β-modification passes back to the α-species. This is accomplished through a fast solid-phase reaction on heating the complex in air at 90°C[238] or keeping it for 24 h in air in the presence of moisture.[241]

The complex composed of $CuCl_2$ and 2-VP was produced in the presence of excess 2-VP in acetone, whereas the mixing of the equimolar amounts of the components in methanol gave rise to weakly precipitating $Cu(2\text{-}VP)Cl_2$ crystals.[238] The amorphous form of this MCM precipitated from ethanol.

The process in which this MCM type is formed is affected substantially by the nature of the complexing agent anion. Thus, the separation of the $Co(4\text{-}VP)_2(NCS)_2$ complex which is similar to $Co(4\text{-}VP)_2Cl$ is impracticable even in that case when the amount of the 4-VP added to the reaction mixture is smaller than a stoichiometric one. In this case $Co(4\text{-}VP)_4(NCS)_2$ is formed.[242]

Among VP complexes with platinum group salts, mention should be made of a $Pd(4\text{-}VP)_2Cl_2$ adduct for which synthesis is conducted in aqueous-ethanol mixtures starting with K_2PdCl_4.[243] Complexes of other metals, namely, Ru(II) and Os(II), are only stable when used in combination with additional chelate-type agents. They will be discussed in Chapter 4.

The VP complexes represent colored crystalline (sometimes amorphous) powders, normally resistant to the oxygen and moisture in the air. Many of them are quite soluble in water and less soluble in alcohols. The $Pd(4\text{-}VP)_2Cl_2$ complex resists the action of any solvent,[242] and 2-VP complexes incorporating Ti(IV), V(III), and V(IV) halides are only slightly soluble.[240]

Note a fairly low heat resistance of $M(4\text{-}VP)_4X_2$ complexes. Thus even at 50°C, $Co(4\text{-}VP)_4(NCS)_2$ starts decomposing: this process terminates at 104°C.[242] A similar chloride complex is still less stable: it decomposes readily to the tetrahedral α-configuration—$Co(4\text{-}VP)_2Cl_2$—even when kept for a long time in air, recrystallized from alcohol, or heated to 113°C.[238] Complexes with two 4-VP molecules display a higher stability. Thus, $Co(4\text{-}VP)_2Cl_2$ decomposes to $Co(4\text{-}VP)_2Cl_2$ only on being heated 90 to 120°C in high vacuum.

B. VINYLAZOLE-BASED MCMs

Vinylazole complexes appear to comprise the best studied class of donor-acceptor-type MCMs. The MX_n is coordinated with vinylimidazoles in nearly all cases at N(3), a pyridine atom of the nitrogen cycle. This has been illustrated using a large variety of methods such as IR spectroscopy,[244-249] electron spin resonance (ESR),[246,250-252] and electron spectroscopy for chemical analysis (ESCA).[250] The situation in this field can be exemplified by the structure of a 1-vinylbenzimidazole (VBI)—$Co(VBI)_2Cl_2$—complex (Figure 13).[253] The N(3) atoms of the two ligands and the chlorine atoms are located at the apexes of a somewhat distorted tetrahedron. The molecule shows the presence of a typically tetrahedral coordination: shortened Co—Cl and Co—N bonds (2.218 and 2.025 Å, respectively). The complexing of VBI and 1-vinylbenzotriazole (VBT) is characterized by a slight shift of the corresponding ligand frequencies, which is attributed to a low basicity of these molecules.[248]

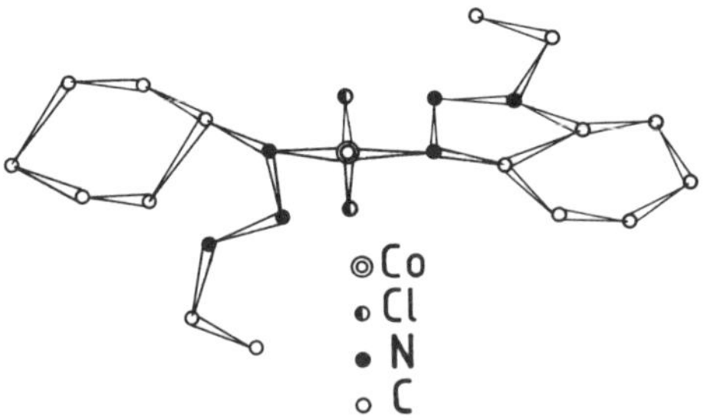

Figure 13 Crystalline structure of $Co(VBI)_2Cl_2$ complex.

The 1-vinylimidazole (VIA) coordination:[254, 255]

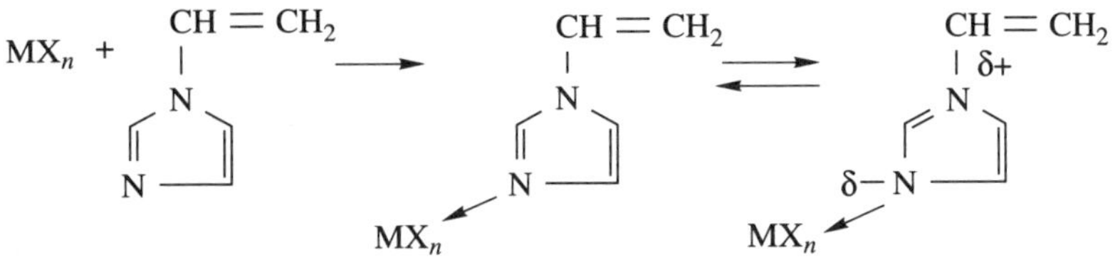

involves a more significant shift of the bond frequency vibrations. Thus, in complexing with Mn(II), Fe(III), Co(II), and Cu(II) chlorides the heterocycle at 1495 and 1511/cm shifts to the shortwave region by 10 to 30/cm.[244] Coordination with Pd(II) results in a high-frequency shift of the 1500 and 870/cm bands by 15 to 20/cm;[256] and in the case of Pt(IV), by 10 to 20/cm.[246] At the same time, $\nu_{C=C}$ (1650/cm)[244,246] or 1652/cm[249] remains virtually the same in terms of both position and intensity.

The complexing of alkyl-substituted VIA runs somewhat differently.[245,247,255] In particular, when coordinated with Mn(II), Fe(III), Co(II), Ni(II), and Cu(II) chlorides, $\nu_{C=C}$ at 1646/cm retains its position with a parallel intensity drop. Still greater is this drop when alkyl-substituted VBI are coordinated with $CoCl_2$.[254] The coordination interaction transfer to the vinyl group in such ligands also has been corroborated by electron spectroscopy.[245]

In ESCA spectra of VIA-based complexes, one can observe a $2p_{3/2}$ metal line shift toward the lower electron binding energy side, 855.8 → 855.2 eV for Ni(II) and 935.6 → 934.1 eV for Cu(II); whereas the same lines for the ligand shift to the higher side.[249] Such a change points to a donor-acceptor type of the M—N bond. The shift for the N(3) lines are more pronounced as compared with the change in the position of the lines for the "pyrrole" N(1) atom: 398.4 → 399.6 eV for Co(II), 399.2 eV for Ni(II), and 399.5 for Cu(II); 400.5 → 400.8 eV, 400.7 eV, and 400.6 eV for the same metals, respectively. A small change in the binding energy of N_{1s}, a "pyrrole" electron atom, might be due to the effects of π-conjugation in the imidazole cycle.

Generally, syntheses of the VIA-based MCMs consist of mixing the solutions of the reaction components in the presence of a solvent such as alcohol, acetone, ether, ether-ethanol composition, or even water.[244,257] The composition of the resulting products is affected by the $[MX_n]$:[L] ratio as well as metal and anion types. Thus, there are reports on the formation of $M(VIA)_6X_2$ complexes, where M = Mn(II), Fe(III), Co(II), Ni(II) or Cu(II); $X = NO_3^-, ClO_4^-$, or BF_4^-.[258,259] On the other hand, for the chlorides of these metals a 1:4 ratio is typical even with excess ligand. At the same time, $CuCl_2$ forms with VIA variously composed complexes (1:4, 1:2, 1:1) according to the ratios between the reagents employed.[244]

The VIA-based donor-acceptor-type MCMs containing Pt(II) are synthesized in an aqueous environment:[249]

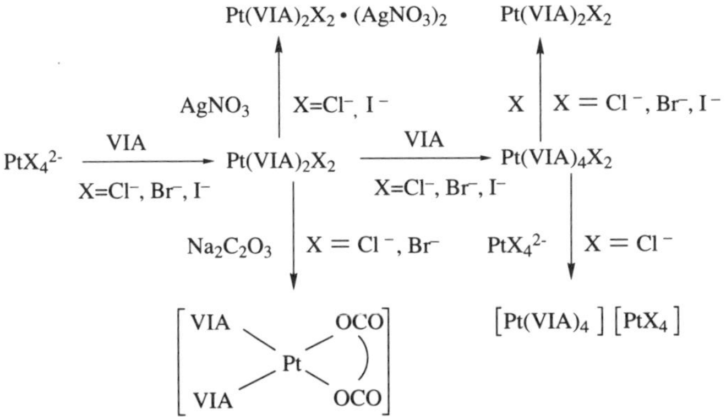

It should be noted that, as indicated by electric conductivity measurements, in water and dimethyl formamide (DMFA) the $[Pt(VIA)_2X_2]$ adducts are nonelectrolytes (μ = 8.17 Ohm/mol·cm^2, whereas $[Pt(VIA)_4]X_2$ belong to three-ion electrolytes (μ = 202 to 255 Ohm/mol·cm^2). The Pt(IV) adducts are synthesized at 20 to 95°C in alcohol, acetone, or water.[246] As a rule, the composition of the resulting complexes corresponds to a 1:2 ratio. As the initial ratio is increased to 1:8 and the temperature to 70 to 95°C, VIA in acetone forms with $PtCl_4$ a polymeric product and in alcohol, a $Pt(VIA)_4Cl_4$ complex.

The complexing of MX_n with other types of vinylazoles is generally characterized by a smaller number of coordinated ligands. Thus although there are products resulting from the complexing of 1-vinyl-2-methylimidazole (ML_4X_2, where X =ClO_4^-, NO_3^-, or Br^-,[258, 260], the ML_2X_2 adducts predominate for the chlorides.[245,247,255] Similarly VBI, VBT, and their alkyl derivatives (irrespective of the ratio between the starting reagents) when reacted with the Mn(II), Fe(III), Co(II), and Ni(II) chlorides, afford ML_2Cl_2 adducts:[244,248,254]

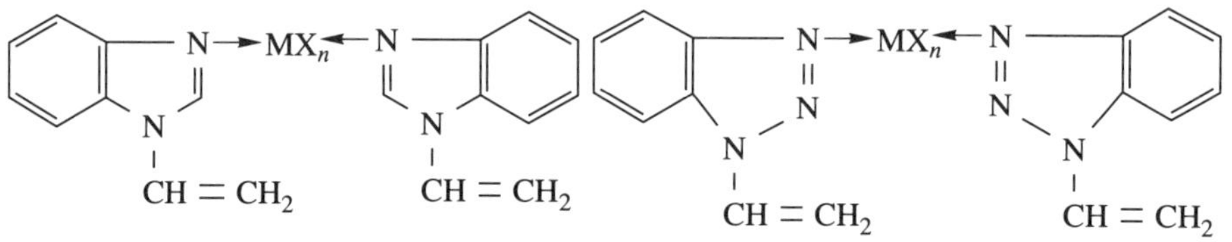

which is due to the steric effect of the condensed benzene ring. Both CuL_2Cl_2 and CuL_4Cl_2 are typical of Cu(II) only.[260]

It has been possible to obtain complexes with other representatives of vinylazoles such as vinylpyrazole, 1-vinyl-3-methylpyrazole, 2-vinyl-1,2,3-triazole, and 1-vinyl-1,2,4-triazole.[256, 261-263] There are also reports on a successful synthesis of vinylcarbazole complex ($VLCl_4$) in which V(IV) was partially reduced to V(III).[264]

Finally, note the production of complex compounds of such transition metals as Pd(II), Rh(III), Ni(II), Co(II), Cu(II), Ti(IV), and V(IV) with 1- and 2-methylvinyltetrazoles.[265, 266] In particular, the PdL_2Cl_2 complex was obtained by stirring the starting reagents in chloroform at 50°C for 24 h on precipitation with petroleum ether.[267] As indicated by X-ray structure studies, the crystals of this MCM are monoclinic and have the following lattice parameters: a = 8.252(3), b = 6.953(2), c = 13.657 (3) Å, γ = 74.93(1)°, Z = 2. The resulting complex is centrosymmetric plane square, Pd(II) being coordinated with the nitrogen atom in position 4 of the tetrazole cycle.

Thus, it has been possible to obtain donor-acceptor-type MCMs with a wide variety of vinyl derivatives of azole—such as carbazole, imidazoles, triazole, pyrazole, and tetrazole—containing one, two, three, and four nitrogens, respectively.

Vinylazole complexes are crystalline substances with colors typical of the corresponding metals and normally stable in air. Most of them are well soluble in water, alcohol, acetone, DMFA, and nitromethane. The ML_6X compounds have an octahedral configuration whereas ML_4X may be in the form of both a tetrahedron, plane square, and even distorted octahedron.[258-260]

C. UNSATURATED NITRILE COMPLEXES

In the series aliphatic amine-nitrogen heterocycle-nitrile the carbon atom hybridization changes from sp^3 to sp with the result that the negative charge gets concentrated at this atom and the nitrogen basicity decreases in this series:[268]

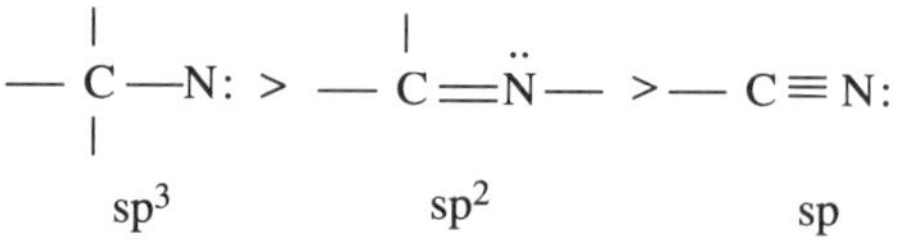

Unsaturated nitriles contain two coordination sites: a lone electron pair at the nitrogen atom and π-electrons of the unsaturated bond. A joint participation of the *p*- and π-electrons in the complexing process is also possible. Examples of all types of such complexes are known. We shall restrict ourselves to the discussion of the *nV*-type complexes.

The interaction of acrylonitrile (AN) with metal salts affects its IR spectrum far more significantly than in the case of VP and vinylazoles in that during the complexing of the Al(III), Zn(II), Ni(II), Ti(IV), V(V), and Pd(II) chlorides the $\nu_{C—N}$ shift to the shortwave region is as high as 37 to 60/cm.[269-274] At the same time, $\nu_{C=C}$ changes by no more than 10 to 15/cm[270,271,275] which can be attributed to an electronic density redistribution in the AN molecule as a result of coordination. There is also a marked drop of $\nu_{M—Cl}$, down to 80/cm.[275] This might be due to the specific features typical of the electron adduct structure. The fact is that in the nitriles the $R—C{\equiv}N \rightleftarrows R—\overset{+}{C}{=}\overset{-}{N}$ structure may resonate; whereas for the complexes, $R—C{\equiv}\overset{+}{N}—\overset{-}{M}Cl_n$ is the most probable structure.[276]

In methacrylonitrile (MAN) complexing with $ZnCl_2$ and $SnCl_4$ $\Delta\delta_{C\equiv N}$ is somewhat lower (15 to 40/cm).[277] There is also a slight low-frequency shift of the $\nu_{C=C}$ band (5/cm). These MCMs are usually synthesized by mixing MX_n and excess ligand in the absence of any solvent. This method gave, in particular, Zn(II),[278] Nb(IV), Ta(IV), Nb(V), and Ta(V)[279] complexes. The composition of the resulting MCM is either MLX_n or ML_2X_n. A new technique has been proposed for synthesizing AN complexes with Al(III), Zn(II), and Ti(IV) compounds at –78°C followed by removal of the free ligand.[272] Another interesting technique suitable for the production of such MCMs consists of the extraction of metal chlorides in Soxlet's apparatus.[280] The resulting complexes then come off by crystallization on further keeping the reaction mixture. This method, together with dissolution in hot AN, has been useful in producing a large number of adducts based on Zn(II), Ti(III), Ti(IV), V(III), V(IV), Cr(III), Mn(II), Co(II), Ni(II), Cu(II), Fe(III), and mixed-valence Fe(II, III) chlorides.

Titanium tetrachloride forms two adducts with AN: a five-coordinated $Ti(AN)Cl_4$ (a tetragonal pyramid) and a six-coordinated $Ti(AN)_2Cl_4$ (an octahedron).[281] The interaction of AN with VCl_4 at 25°C gives—depending on the ratio of the starting reagents—$V(AN)_nCl_4$ complexes, where $n = 1$ to 4; the highest stability is displayed by the species with $n = 2$ and 3.[270, 271] As reported by other researchers, the $V(AN)_2Cl_4$ complex is only formed at a reaction temperature of –70°C in the dark.[279] At the same time, $VOCl_3$-based MCMs give no stable AN complexes at all.[270, 271, 281]

Among complexes based on platinum group metals, one can mention only the synthesis of an Ir(I)-containing adduct:[282]

$$Ir(ClO_4)(CO)[(C_6H_5)_3P]_2 + AN \rightarrow Ir(AN)CO[(C_6H_5)_3P]_2ClO_4$$

D. CYCLIC AMINE COMPLEXES

Incorporation of cyclic amine complexes into donor-acceptor-type MCMs is justified by their ability, in the appropriate conditions, for cycle opening followed by formation of metal-containing polymers.

Complexing with ethyleneimine, its N-substituted species as well as with triethylene imine N-derivatives, is mainly accomplished with the use of $ZnCl_2$ and sometimes Cd(II), Co(II), Ni(II), and Pd(II) chlorides.[283-291] Coordination occurs at the nitrogen atom, the $\nu_{C—H}$ and the bending vibration of the cycle being shifted by 20 to 39/cm to the long-wave side. There are also shifts of the C—H vibration absorption band together with a noticeable drop of the electron density at the C—N bond. This suggests that formation of a donor-acceptor bond greatly affects the electronic structure of small heterocycles.

Imine complexes are synthesized in the presence of methyl and ethyl alcohols or ether at room or lower temperature (–10°C). Ethylene imine complexes have the following compositions:

$$M(\overline{HN-CH_2-CH_2})_nX_2$$

where n = 1 to 4; whereas ethylene- and triethylene imine derivatives give adducts of the same type:

$$Zn(\overline{RN-CH_2-CH_2})_2Cl_2$$

and

$$Zn(\overline{HN-CH_2-CH_2-CH_2})_2Cl_2$$

respectively, where R = $C_6H_5CH_2CH_2$, and others. Most of these complexes are white crystal powders with a fairly low melting temperature (56 to 90°C).

E. COMPLEXES BASED ON OXYGEN-CONTAINING LIGANDS

Unlike polymerizable complexes with nitrogen-containing ligands, their oxygen-containing counterparts have been little studied. The following monomers were used for complexing: methacrylate (MA), methyl methacrylate (MMA), and vinyl acetate (VA). It has been possible to obtain $TiCl_4$ complexes with MA (1:1 and 1:2).[275] As compared with a free complex, the $\nu_{C=O}$ and $\nu_{C=C}$ were found to be 85 and 60/cm lower, respectively (in MA the corresponding values are 1725 and 1632/cm respectively). However, the complexing runs at the carbonyl group, which is corroborated by a high dipole moment of the complex (5.48 D). Note that in the π-complexes the dipole moments are far lower. In the course of adduct formation some other changes in the IR spectra of MA are also observed: disappearance of the ester band at 1200/cm and emergence of a 990 to 975/cm doublet instead of an out-of-plane $CH_2=CH—$ mode. Thus although the complex is formed at the carbonyl group, all the bonds in the molecule are affected by the electron cloud shift.

Coordination of MMA also leads to a noticeable $\nu_{C=O}$ decrease (by 65 to 130/cm).[272] However, in this case the $\nu_{C=C}$ changes are relatively small.[264, 292, 293] At the same time, the ester group absorption bands show a strong shift to the high-frequency region (1168 → 1302, 1250 → 1340/cm). The magnetic susceptibility and ESR evidence suggest that vanadium in $V(MMA)Cl_4$ complexes is contained in the V(IV) form.

Also, nuclear magnetic resonance (NMR) studies of the complexes:

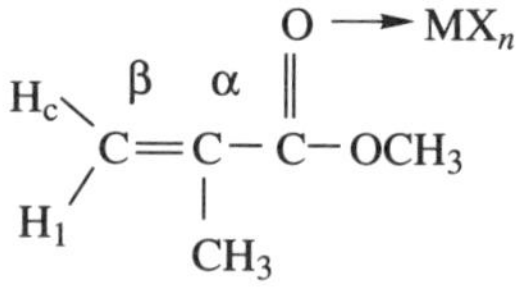

(MX_n = $SnCl_4$, $TiCl_4$, $AlCl_3$, etc.) indicate that the chemical shifts of all the protons in the bonded ligand are far more pronounced as compared with those in the free one.[272] Also, the vinyl *trans*-proton shift relative to the ester group (δ_{H_1}) increases with the Lewis MX_n acidity. This is coupled with a decrease in the electron density at the carbon atom in the β-position. In most cases MMA complexes (MLX_n) have been isolated, whereas ML_2Cl_4 adducts have been obtained with $SnCl_4$ and $TiCl_4$ only.[275,292]

In the case of VA the only products obtained were the equimolar adducts of Ti(IV) and V(IV) chlorides with the structure close to that of MMA complexes.[282]

Synthesis of MCMs with oxygen-containing ligands is run in both the presence and (more often) in the absence of solvents. Note that the process during which $Zn(MMA)Cl_2$ is produced should be conducted in the dark, and the MMA excess should be removed *in vacuo* considering the high tendency of this MCM to polymerization.[293] A synthesis in the solid phase and at low temperature has been suggested.[295] It has been useful in the production of a $Ti(MA)Cl_4$ as well as a ternary adduct $TiCl_4$—MA-cyclopentadiene, with a somewhat unclear structure.

F. COMPLEXES OF AAm AND ITS DERIVATIVES

It is known that acrylamide (AAm) is a special kind of ligand since its molecule contains simultaneously two *p*-donor groups, namely, a carbonyl and an amine one which forms an amide group. In its N-derivatives an atom (or both atoms) of the hydrogen at the NH_2 group is (are) substituted by an alkyl residue. The complexing is effected using—along with AAm—*N-tert*-alkyl derivatives such as butyl, amyl, hexyl, and heptyl; *N,N*-dimethylacrylamide; methacrylamide; and *N-tert*-butylmethacrylamide.

For the ligands of this type it is very essential where the coordination bond is localized. The existence of the double bond in AAm markedly affects the electronic state of the amide group. Assuming the resonance of the C=C and C=O groups with the participation of NH_2, the formation of three differing resonance structures is deemed likely:[296]

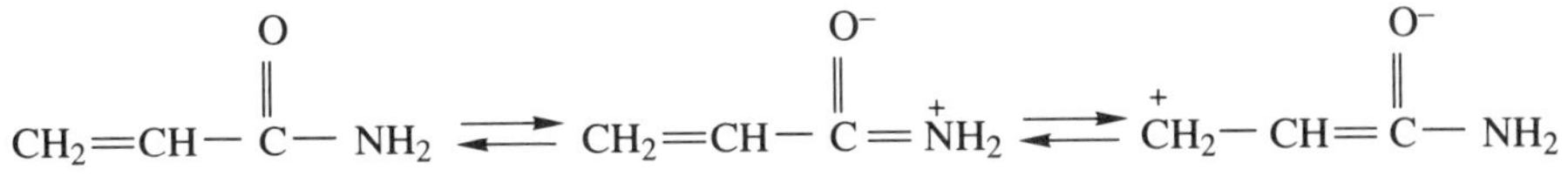

I II III

In complexing via the oxygen atom structure II is preferred, whereas structures I and III are expected to exist in the case of N-coordination.

The presence of such resonance AAm structures seems to be responsible for the fact that during complexing one can observe frequency changes of not only the stretching and bending modes of the C=O, C—N, and N—H bonds but also those of the vinyl group. As this takes place, in most cases these shifts that may be as high as 40 to 50/cm are oriented toward the low-frequency region.

A parallel shift of the ν_{NH}, ν_{CO}, and δ_{NH} bands to the long-wave region—caused by ligand complexing—gives rise to conflicting conclusions as to MX_n coordination. Some authors believe that complexing takes place at the nitrogen atom,[296-298] and others advocate the same process occurring at the carbonyl oxygen as in the case of saturated amines.[299-302]

Nevertheless, X-ray structure studies of the $Co(AAm)_4(H_2O)_2(NO_3)_2$ complex enable one to think that AAm is coordinated via the oxygen atom (Figure 14).[301] The MCM structure is composed of octahedral $[Co(AAm)_4(H_2O)_2]^{2+}$ cations and NO_3^- anions united through a three-dimensional system of hydrogen bonds. The spread in the Co—O distances (2.065 to 2.114 Å) suggests ligand nonequivalency. Also, the double ligand bonds are different, 1.298 and 1.318 Å for C(15)—C(16) and C(12)—C(13), respectively; this might be due to the participation of C(12) and C(16) in the intermolecular interactions. A distinctly shortened C(13)··· C(13)′ contact equal to 3.23 Å is matched with a longer C(12)—C(13) bond. In other words, this bond can be regarded as the existing "workpiece" ready for polymerization. The conclusion that the metal is coordinated via the AAm oxygen atom is additionally confirmed by electron spectroscopy of the $M(AAm)_4(H_2O)(NO_3)_2$ and $M(AAm)_4Cl_4$ complexes, where M = Co(II) or Ni(II).[303, 304]

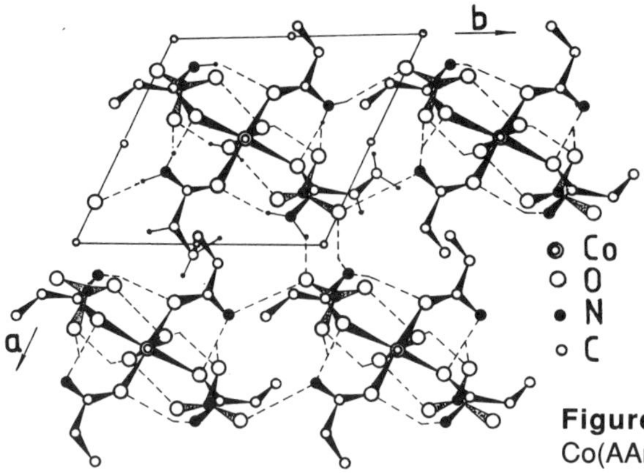

Figure 14 Crystalline structure of $Co(AAm)_4(H_2O)_2(NO_3)_2$ complex.

The results of studies into the structure of N-substituted AAm derivatives are more definitive, suggesting that the presence of substituents is bound to handicap the complexing at the nitrogen atom. In fact, mass spectrometric evidence indicates that the coordination bond of Sn(IV) with *N,N*-dimethacrylamide is formed at the oxygen atom.[298] The IR spectrum of the complex starts showing a new absorption band, namely, Sn—O. Coordination of other AAm derivatives such as *N-tert*-butyl-acrylamide and *N-tert*-amyl-acrylamide with $ZnCl_2$ in dioxane involves a shift of the vinyl group quartet center in the NMR spectrum toward the low field and a decrease in the spin-spin interaction constant.[305]

It is known that MX_n compounds form with AAm complexes having various degrees of coordination. They can be produced using techniques similar to those described above for VP, VIA, and AN (e.g., in acetone or alcohol solutions with ether precipitation). However, the heaviest reliance has been put on a method consisting of grinding AAm together with either a crystalline MX_n hydrate or anhydrous salts containing small amounts of water, followed by washing in hot benzene and ether and drying. Thus, it has been possible to obtain adducts and chlorides of Zn(II), Co(II), Ni(II), Cu(II), Mn(II), Cr(III), and Fe(III).[297,301,302,306] The composition of all chloride complexes as well as Cu(II) and Fe(III) nitrates was found to correspond to the formula $M(AAm)_4X_n$; whereas the adducts of Mn(II), Co(II), Ni(II), and Zn(II) nitrates, $[M(AAm)_4(H_2O)_2(NO_3)_2]$, incorporate crystal water. It is interesting that AAm complexes of metal nitrates as well as of Ni(II) chloride have a pseudo-octahedral structure, whereas $Co(AAm)_4Cl_2$ changes its structure from a tetragonally distorted octahedron in crystal to a tetragonal one in solution.[303, 304]

Similar approaches have been applied to the synthesis of AAm derivative complexes. Thus, adducts of various *N*-alkylacrylamides with Zn(II), Cd(II), and Hg(II) halides were synthesized using the following three techniques: (1) reaction in anhydrous ether followed by precipitation of the complex obtained, (2) dissolution of the components in a minimum amount of boiling ethyl alcohol followed by a slow evaporation of the boiling solvent at reduced pressure, and (3) mixing in a mortar in the presence of water traces followed by drying and rinsing with warm benzene.[306] The composition of the resulting complexes was ML_2X_2.

In the case of AAm itself the similarly composed complexes were also obtained for Zn(II), Cd(II), and Sn(IV) salts; the complexing was performed in alcohol, ether, and chloroform.[296,306] At the same time, in ethanol solutions of triethyl orthoformiate and 2,2-dimethoxypropane (dehydrating agents) at a sevenfold molar excess of AAm the crystalline hydrates of some Zn(II), Cd(II), Mn(II), Fe(II), Co(II), Ni(II), and Cu(II) salts gave $M(AAm)_6X_2$ complexes.[307,308] However, nitrates of Mn(II), Cu(II), and Cd(II)—as distinct from their perchlorates—do not give such adducts even at a large ligand excess. The largest coordination number (CN) is also observed in the complexing of $Co(ClO_4)_2 \cdot 6H_2O$ with AAm in nitromethane.[299] It should be noted that the complexing ability depends markedly on the type of not only the MX_n anion but also on that of the metal. Thus, $NiSO_4$ is absolutely incapable of complexing with AAm.[309]

Complexes of AAm and its derivatives comprise crystal powders well soluble in water, alcohol, and acetone; and insoluble in benzene, ether, and other nonpolar species.

G. MCMs BASED ON SULFUR- AND PHOSPHORUS-CONTAINING LIGANDS

MCMs of this type are still rather scarce. Thus, it has been possible to produce complexes with some alkylvinyl sulfooxides, described by the general formula:

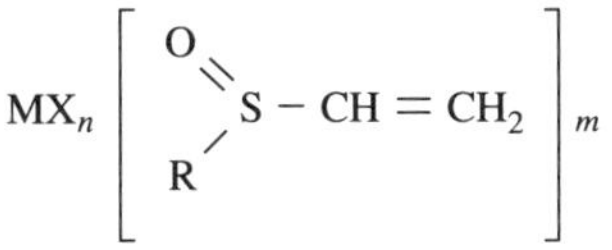

where M = Co(II), Mn(II), Cu(II), Ni(II), Cr(III), or Fe(III); X = Cl^- or NO_3^-; R = C_4H_9 or C_8H_{17}; n = 2 or 3; m = 2, 3, or 4.[310,311]

The IR spectra of the complexes obtained show the shifts and broadening of the $\nu_{C=C}$ band from 1600 to 1610–1640/cm. In the NMR spectra of the $ZnCl_2$ vinyl sulfoxide complex the double bond proton signals shift toward the low field to $\Delta\delta$ = 1 ppm, and in the NMR ^{13}C spectrum one can observe signals of the α- and β-carbon atoms of the groups coming closer together. Based on these data, the authors of the work cited above were led to believe that the vinyl group is directly involved in the complexing process. However, such facts might be merely attributed to the electronic density redistribution in the ligand molecule due to the coordination bonding; this has been stated for the formation of MCMs containing AN, MMA, and other substances.

Note that some MCMs of this type comprising butylvinyl sulfoxide with Fe(III), Cr(III), and Co(III) nitrates $[ML_3(NO_3)_3]$ are very unstable.[311] They can undergo spontaneous explosive decomposition even at the stage of synthesis or subsequent storing with the appearance of nitrogen, sulfur, and metal oxides as well as coke like products. Among these species the Fe(III) complex is the least stable. It explodes at 20 to 40°C both individually and in various solutions at concentrations over 40 mol%. It is interesting that complexes of divalent metals and trivalent chlorides are quite stable. Obviously, this phenomenon can be accounted for by thermal explosion due to a spontaneous polymerization of space-arranged monomeric "workpieces" resulting from the self-association of the vinyl sulfoxide complexes of trivalent metal nitrates due to the Van der Waals and coordination interactions.

MCMs based on phosphorus-containing ligands are produced with the use of phosphines incorporating unsaturated groups such as vinyl and styryl ones. The Co(II), Ni(II), and Pd(II) complexes with styryl phosphines are obtained through either direct interaction of the reagents or ligand exchange with complexes containing substitutable ligands such as phosphines, nitrile, and acetyl acetonates:[312]

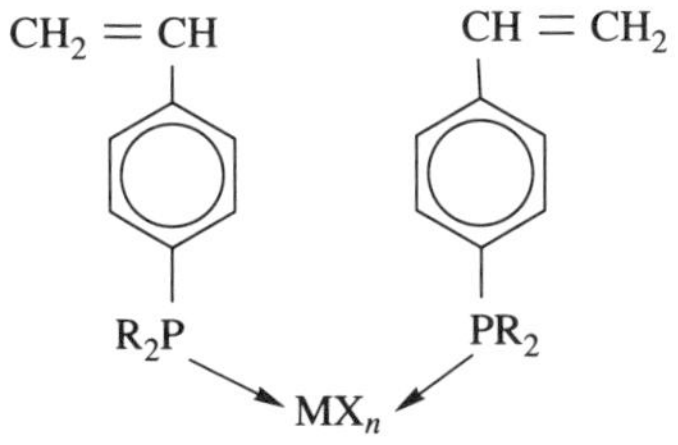

A similar Pt(IV)-containing MCM was obtained differently:[313]

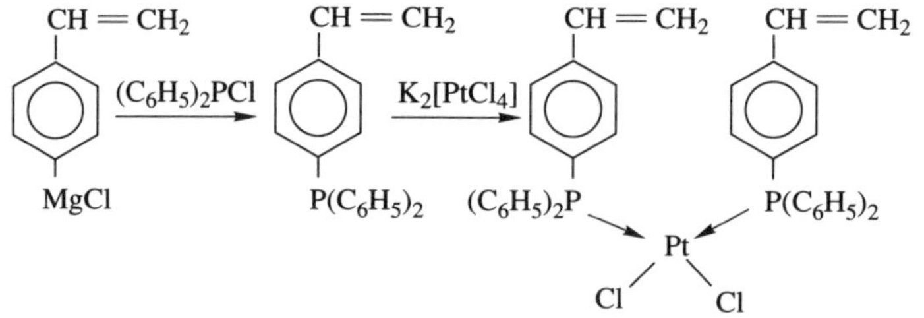

Mention also can be made of how complexes with the same ligand and Cr(VI), Mo(VI), and W(VI) carbonyl derivatives are synthesized.[314,315] A somewhat more complicated technique was used to obtain allyl MCMs of this type:[315]

$$Cr(CO)_4\left(P(C_6H_5)_2H\right)_2 \xrightarrow{C_4H_9Li} \xrightarrow{CH_2=CHCH_2Cl} \left[Cr(CO)_4\left(P(C_6H_5)_2H\right)\left(P(C_6H_5)_2\right)\right]$$

$$\rightarrow iso-(CO)_4Cr\left(P(C_6H_5)_2H\right)\left(P(C_6H_5)_2CH_2CH=CH_2\right)$$

II. HOMOPOLYMERIZATION OF DONOR-ACCEPTOR-TYPE MCMs

Polymerization of this type of MCM is based on a very important assumption by which the composition of the resulting polymer corresponds to that of the starting monomer. Polymerization of 4-VP complexes can be exemplified by the following scheme:

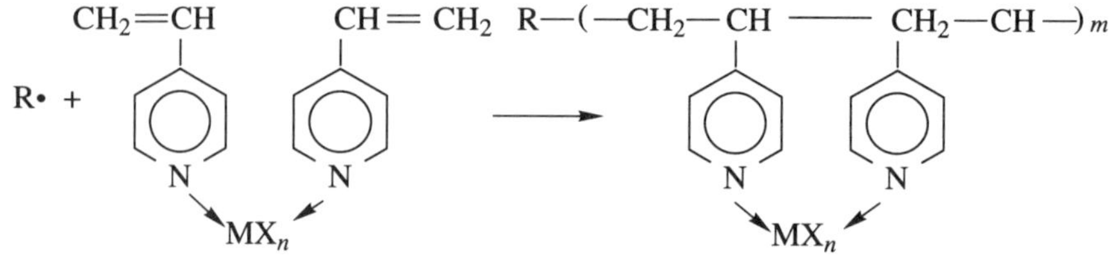

Note that such metal-containing polymers can be also obtained differently, i.e., using polymer-analogous conversions of the corresponding macroligands with MX_n:

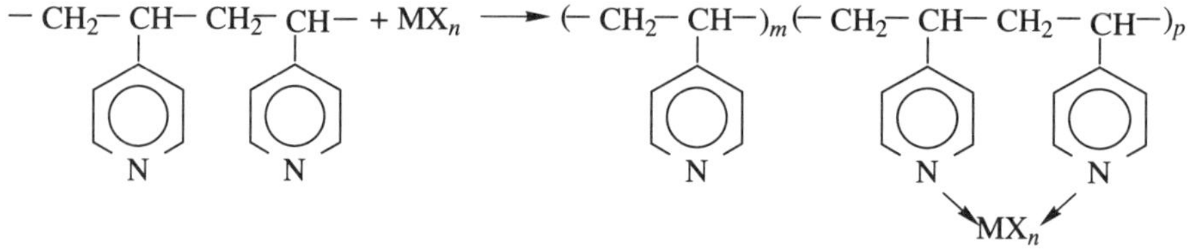

However, such polymers are distinguished by high composition inhomogeneities in that not all the functional groups of the starting macromolecule interact with MX_n for several reasons (changes in the conformation and shape of the reacting macromolecule, "the neighbor effect", etc.). At the same time, as noted in the introduction of this chapter, the specificity of MCM polymerization conversions is just what is related to the structural homogeneity of the resulting macrocomplexes with no noninteracting functional groups therein.

Donor-acceptor-type MCMs are synthesized using a variety of experimental approaches, both conventional and comparatively new.

A. POLYMERIZATION IN SOLUTIONS

The first studies that illustrate the effect of complexing additives on radical polymerization of conventional monomers date back to the late 1950s.[316–318] They marked the starting point for looking into polymerization behavior of *nV*-type MCMs. Polymerization of such MCMs offers the possibility of changing their composition in solution as a result of dissociation, entering the solvent molecules into the coordination sphere of the metal complexer, etc. However, in most cases dissociation is relatively insignificant and the complexes under study give rise to polymers with a composition that is similar to that of the parent MCM. Such MCMs include, among others, Zn(II) adducts with 4-VP, 2-VP, and MVP.[239] Their polymerization in various solvents suggests that the complexing of VP (except for 2-VP) leads to a noticeable (by 1.5 to 5 times) process acceleration as compared with the case when an uncoordinated ligand is used. In acetone, tetrahydrofuran (THF), or ethanol, polymerization is heterogeneous even at the early stages. In the case of ethanol one can observe the effect of self-acceleration. The polymerization rate of complexes with different ligands is markedly dependent on the vinyl group position in the ring:

$$Zn(4\text{-}VP)_2(SCN)_2 > Zn(MVP)_2(SCN)_2 > MVP > Zn(2\text{-}VP)_2(SCN)_2$$

The nature of the anion affecting, to a large measure, the Lewis MX_n acidity is also important in that the rate of the MVP complex polymerization in DMFA (in homogeneous conditions) at 70°C decreases in the order $SCN^- > Cl^- > Br^- > I^-$. The effects observed seem to stem from the totality of factors including a steric double bond accessibility. The molecular mass (MM) of the growing polymers increases linearly with the MCM polymerization rate. At the same time, in the case of MVP complex polymerization, the product MM changes only slightly, within $(1 \text{ to } 3)\cdot 10^4$ (for the polymer synthesized from the individual MVP, MM = $1\cdot 10^4$).

Note that other types of the Zr(II) VP-complex polymerization initiation are less effective as compared with the radical ones. Thus, the anionic C_4H_9Li initiator in THF at 25 to 60°C provides for polymer yield not exceeding 5%, which is attributed to a significant contribution of the side reaction between the initiator and ZnX_2. Cationic initiation ($SnCl_4$, THF, at temperatures between –78 and +40°C) is ineffective due to the appreciable positive values of the parameters *e* in such MCMs.

Among VP transition metal complexes mention should be made of adducts composed of $NiCl_2$ ($NiBr_2$) and 4-VP, effectively copolymerizable in DMFA at 100 to 120°C in the presence of azobisisobutyronitrile (AIBN).[319]

Despite the existence of advanced methods suitable for synthesizing MX_n complexes with vinylazoles and a large amount of experimental material to this effect, evidence concerning their polymerization conversions is much more modest. It should be noted that homopolymerization of some of these MCMs involves an appreciable loss of MX_n as compared with the parent MCM. Thus in the case of the $Zn(VBI)_2Cl_2$ complex, the $ZnCl_2$ content—when passing from the MCM to the polymer—decreases from 31.7 to 28.9%; and the drop of the $Zn(VBT)_2Cl_2$ content is still higher, from 32.1 to 16.2%.[320] The same metal compound depletion is also typical of $MnCl_2$ vinylazole complexes.[321] These facts were attributed to a significant dissociation of the MCM under discussion in DMFA; another possible reason might be steric hindrances during macrochaining.

Fairly unusual "order" dependences of the process rate on initiator and monomer concentrations have been found (Table 13). An enhanced order with respect to the monomer and a decreased activation energy (E_a) for $Zn(VBT)_2Cl_2$, as well as a higher

Table 13 **Kinetic parameters for polymerization of Zn(II) or Mn(II) chloride-vinylazole complexes**[321]

Monomer	Reaction order: With respect to monomer	Reaction order: With respect to initiator	E_a (kJ/mol)	Elementary stage rates: $w_i \cdot 10^7$ mol/l·s	Elementary stage rates: $w_p \cdot 10^5$ mol/l·s
VBI	1.0	0.57	108.0	0.689	1.11
$Zn(VIA)_2Cl_2$	1.2	0.7	85.0		
$Zn(VBI)_2Cl_2$	1.0	1.0	71.5		
$Mn(VBI)_2Cl_2$	1.0	0.34	72.0	2.45	1.34
$Zn(VBT)_2Cl_2$	2.0	0.8	55.4		

order with respect to the initiator for $Mn(VBI)_2Cl_2$, are due to a spontaneous homopolymerization occurring in these systems for which rate is commensurate with the degree of initiation. Spontaneous polymerization was also observed in the case of some other complexes such as $Zn(VIA)_2Cl_2$, $Zn(VBI)_2Cl_2$, and $Mn(VBT)Cl_2$. This polymerization is of the radical type, which is corroborated by the reaction being inhibited with *n*-benzoquinone. It is interesting that this process discontinues after the monomer conversion has been fixed at ca. 15%. For $MnCl_2$ complexes a spontaneous polymerization mechanism has been put forward consisting of a partial oxidation of Mn(II) to Mn(III). The latter, vinylazole-coordinated, species generates radicals responsible for initiating the process.[321] The existence of such a mechanism is supported by the fact that the reaction increases as the system is saturated with oxygen in the air (Figure 15), whereas in the absence of oxygen no spontaneous polymerization is observed.

The rates of AIBN-initiated homopolymerization of the $MnCl_2$ with various vinylazoles without any contribution of the spontaneous process have been compared (VIA > VBI > VBT).[321] The $MnCl_2$-VIA polymerization has been found to be distinguished by the dependence of the polymerization ability on metal CN due to a differing number of the complex-bonded VIA molecules in the MCM. Thus, a comparison of the rates of $Mn(VIA)_2Cl_2$ and $Mn(VIA)_4Cl_2$ polymerization and that of the uncoordinated ligand indicates that the conversion of the former in aqueous methanol is much higher than that

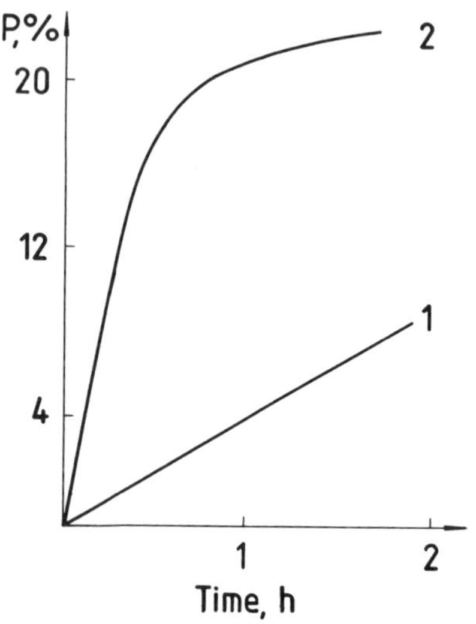

Figure 15 Kinetic curves for spontaneous $Mn(VBI)_2Cl_2$ polymerization: solution of complex in DMFA in air (1), upon additional reaction mixture saturation with oxygen in the air (2).

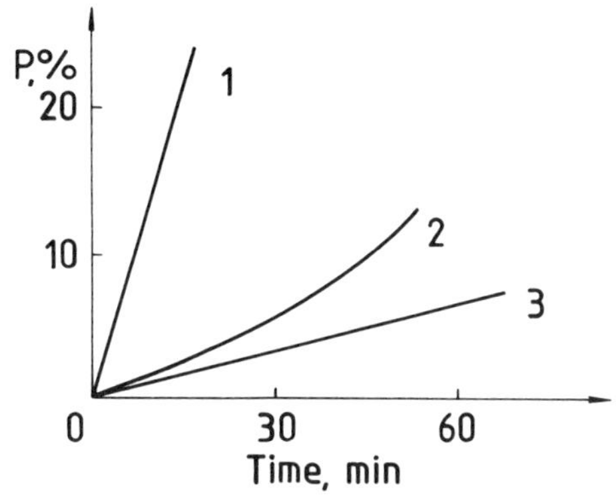

Figure 16 Kinetic curves for complex-bonded VIA polymerization: $Mn(VIA)_4Cl_2$, 0.5 mol/l (1); uncoordinated VIA, 2.0 mol/l (2); VIA·HCl, 0.5 mol/l (3); aqueous methanol, 60°C; [AIBN] = $8 \cdot 10^{-2}$ mol/l.

of VIA (Figure 16).[322,323] In terms of the complexing effect, $MnCl_2$ proved to be more influential than even HCl or CH_3I. It is probable that due to a strong dynamic conjugation of the vinyl group and heterocycle π-electronic systems the coordination with $MnCl_2$ leads to a higher polarity of the C—C bond and a greater parameter *e*. The complexing gives rise to larger $k_p/k_t^{1/2}$ ratios (0.116 and 0.348 $mol^{1/2}s^{1/2}$ for VIA and $Mn(VIA)_2Cl_2$, respectively), which is likely due to a smaller k_t value as a result of a lower molecular mobility of the growing complex-bonded polymer chains.

However, the above properties did make themselves felt in the polymerization of $Mn(VIA)_4Cl_2$ and similar Co(II), Ni(II), and Zn(II) complexes.[323] These MCMs are less active in radical polymerization in alcohol as compared with uncoordinated ligands (Figure 17). In comparable conditions polymer yield in the polymerization of VBI and VBT complexes (ML_2Cl_2) was still lower. Attempts to convert complexes composed of VIA and Fe(III) or Cu(II) to polymers gave no results at all, possibly for the same reasons as in the case of ionic-type MCMs containing these metals.

Special mention should be made of an approach based on the joint synthesis and polymerization of vinylazole complexes.[324] Thus, it has been possible to obtain polymers based on VBI–$MnCl_2$ (4:1), VIA–$NiCl_2$ (4:1), VIA–$ZnCl_2$ (2:1), VBI–$NiCl_2$ (3:1), and VBT–$NiCl_2$ (2:1) systems in alcohol environments. Studies of the effect of the nature of vinylazole on the process indicate that VIA is the most influential species in that polymer yield is as high as 25 to 41% compared with 17 to 21% for VBI and VBT complexes prepared *in situ*.

A platinum $Pt(VIA)_4Cl_2$ complex was shown to polymerize in DMFA (AIBN initiation) to give water-soluble products.[325]

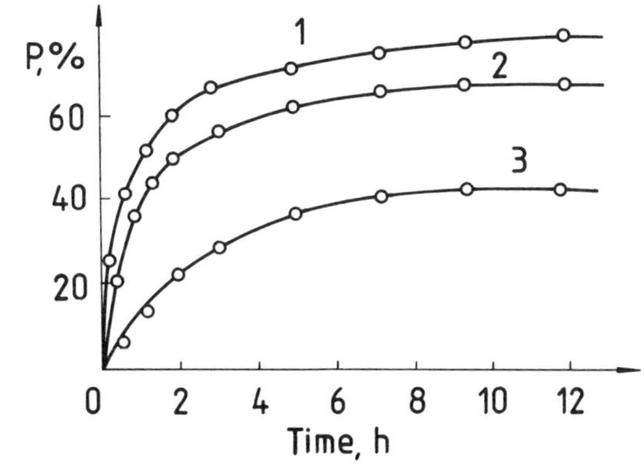

Figure 17 Kinetic curves for polymerization of VIA and its complexes: VIA (1), $Mn(VIA)_4Cl_2$ (2), $Ni(VIA)_4Cl_2$ (3); ethanol, 70°C, 4% AIBN.

Radical polymerization of Co(II) complexes with 1- and 2-methylvinyltetrazoles accompanied by formation of a product containing only a few percent of metal has been studied.[265] It is obvious that in this case, as with vinylazole complexes, a key role is played by dissociation of the MCMs and the metal-containing polymers.

Homopolymerization of AN complexes with Zn(II) halides such as $Zn(AN)X_2$ and $Zn(AN)_2X_2$ in benzene was induced by radiation.[326] A higher polymerization rate was observed for equimolar adducts; and in the temperature range between –78 and +30°C, the polymer yield was found to increase from 5 to 36%, G_{-MCM} = 3.7 to 25 I/100 eV, MM ~ $1 \cdot 10^6$. At the same time, the polymerization of the $Zn(AN)_2X_2$ complexes was distinguished by lower yields of the resulting product and its MM (ca. $6 \cdot 10^5$). It is interesting that addition of polar substances such as water, methanol, and acetic acid to the polymerizable system increases the polymerization rate, on the one hand; and brings about a significant (up to 43%) degree of complex dissociation, on the other hand. In this case, the higher polymerization rate might be due not only to the additional contribution of the initiating particles but also to the changing nature of the process itself, i.e., copolymerization of the free and complex-bonded AN.

In the polymerization of MAN complexes of Zn(II) and Sn(IV) in benzene or toluene (AIBN or radiation initiation), the dissociation played no significant role.[277] The resulting product had a high MM value ($\overline{P}$= 1,600 to 14,000). The polymerization rate for the MAN complexes decreased in the series $Zn(MAN)Cl_2 > Zn(MAN)_2Cl_2 > Sn(MAN)_2Cl_4 > Sn(MAN)Cl_4$ to reach 90% conversion for the first of them. However, in polar solvents (THF-type) a partial dissociation of the complexes was still observed to result in a marked drop of the polymerization rate.

Polymerization of the AN complexes is heavily affected by the type of the complexing metal. Thus for AN polymerization in the presence of various metal chlorides, the activities were as follows:[327]

$$CuCl_2 > CdCl_2 > \text{free AN} > CoCl_2 > NiCl_2$$

Other researchers have copolymerized MMA complexes with Zn(II), Co(II), Ni(II), and Cu(II) chlorides—$M(MMA)Cl_2$—(boiling in DMFA, AIBN initiation).[328] The resulting polymers were insoluble in organic solvents and the metal content (25.6 to 27.7 mol%) corresponded to that in the parent MCM. The rate of the process changed in the series Cu(II) > Zn(II) > Co(II) > Ni(II) > uncoordinated MMA. As exemplified by the $Zn(MMA)Cl_2$ complex, the E_a value of the process reduces by half as compared with the individual MMA.

It has been noted that some donor-acceptor-type MCMs (butylvinyl sulfoxide and vinylazole complexes) are able to undergo spontaneous polymerization both individually and in solutions. Also, in aqueous concentrated solutions AAm nitrate complexes of a number of metals, such as Cr(III), Bi(III), Pb(II), Ca(II), UO_2(II), Th(IV), In(III), etc. (nearly all lanthanides), are capable of spontaneous polymerization at room temperature at the synthesis stage.[329] As this takes place, the reaction mixture-contained water is retained in the resulting polymer. As shown by calorimetry, the way the process runs depends very much on the reaction mixture composition.[330] It is evident from Figure 18 that the kinetic polymerization curves at various [AAM]:[$Cr(NO_3)_3 \cdot 9H_2O$] ratios are mostly very much the same, the only exception to this being the 5:1 ratio. In this case the process occurs at a far higher initial rate followed by fast deceleration. A different dependence was observed when using Er(III) nitrate as the complexing agent (Figure 19). In the latter case, a higher viscosity of the system seems to become the governing factor as the [AAm]:Er(III) nitrate ratio increases.

It should be noted that heat release during the spontaneous polymerization in the AAm-$Cr(NO_3)_3 \cdot 9H_2O$ system at 20°C, 3 to 6 h after the onset of this process, corresponds

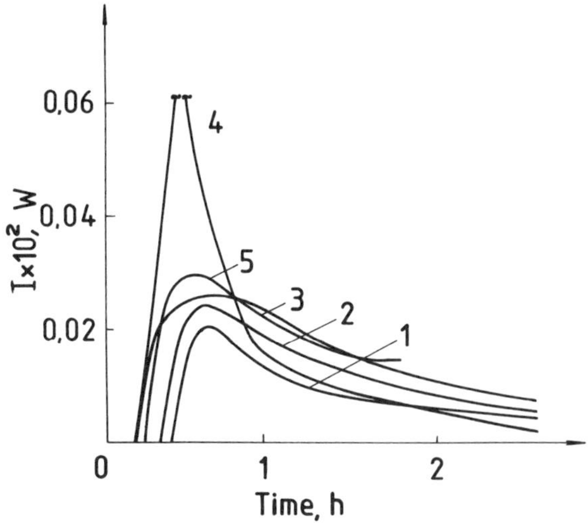

Figure 18 Heat release rate as a function of AAm spontaneous polymerization duration in the presence of $Cr(NO_3)_3 \cdot 9H_2O$ at various reaction mixture compositions, [AAm]:[$Cr(NO_3)_3 \cdot 9H_2O$], molar ratio: 2.0 (1), 3.0 (2), 4.0 (3), 5.0 (4), and 6.0 (5); no water additives, 20°C.

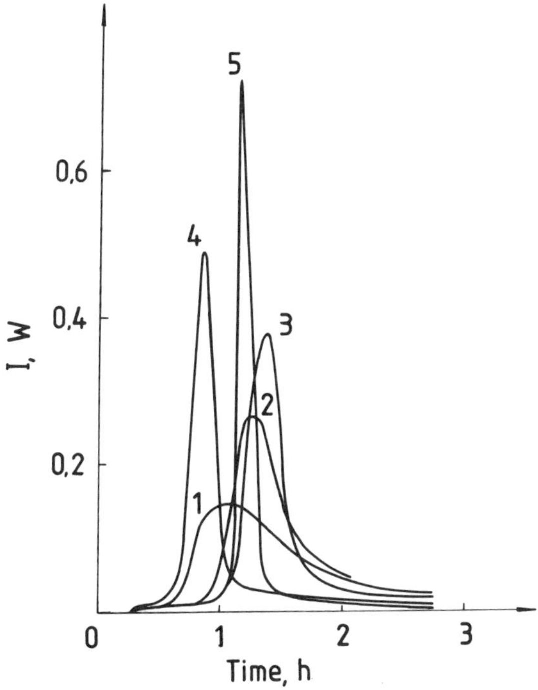

Figure 19 Heat release rate as a function of spontaneous polymerization of AAm complex-bonded to Er(III) nitrate at various reaction mixture compositions, [AAm]:[$Er(NO_3)_3 \cdot 6H_2O$]; molar ratio: 3.2 (1), 6 (2), 8 (3), 12 (4), 20 (5); m_{AAm} = 0.516 g; m_{H_2O} = 0.1 ml; 25°C.

to no more than 20 to 25% of the total heat of AAm polymerization. This might be due to the fact that at the initial stage the reaction involves a comparatively small portion of AAm molecules. The process comes to an end in the solid phase. The second stage becomes evident with increasing temperature (Figure 20). At 60°C the process terminates in ca. 2 h. At the same time, when use was made of other metal nitrates as the complexing agents, polymerization occurred in one stage. It was fairly unexpected to come across the dependence of the spontaneous polymerization rate on the amount of the water added to the system in that there was a shift in the time the reaction commenced and a decrease in its initial rate. Thus, for the spontaneous polymerization of the complex-bonded AAm

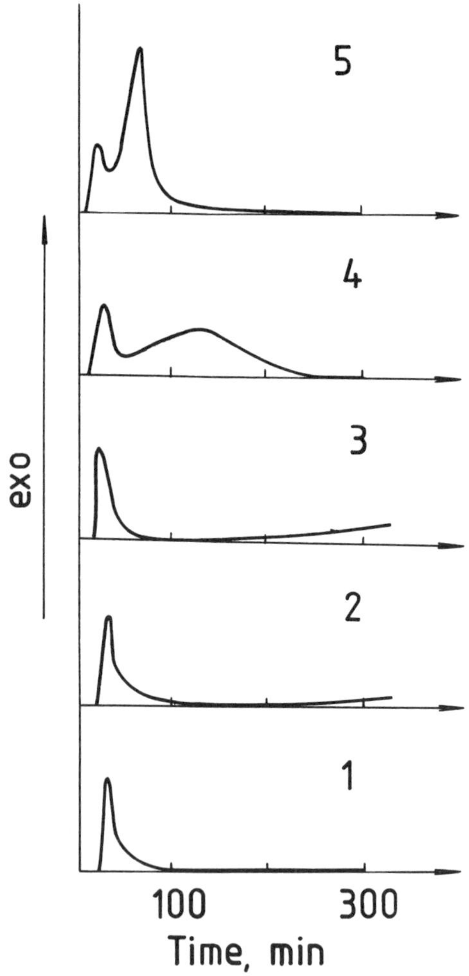

Figure 20 Spontaneous polymerization kinetics of AAm complex-bonded to Cr(III) nitrate (5:1 molar ratio); temperature °C: 20 (1), 30 (2), 40 (3), 50 (4), and 60 (5).

to take place, it is important that the complexes to be polymerized should be sterically close to one another.

In studies of the properties of products resulting from spontaneous polymerization of the complex-bonded AAm, it was found that when kept in water at room temperature a soluble and an insoluble moiety are obtained, the insoluble moiety comprising polyacrylamide (PAAm) which contains a negligible amount of metal compound. In these conditions the process was assumed to give rise to a superpolymer.

The mechanism describing the spontaneous polymerization of AAm that is complex-bonded to metal nitrates is still unclear. However, its occurrence in the case of nitrate salts only enables one to assume that the nitrate groups are involved in a radical-type process.

Some MCMs based on phosphorus-containing styrene derivatives were also found to be capable of radical polymerization. Among these MCMs are $Co[(C_2H_5)_2PCH_6H_4CH{=}CH_2]_2Br_2$, $Co[(C_6H_5)_2(PC_6H_4)_2CH{=}CH_2]_2Br_2$, $Ni[(C_2H_5)_2PC_6H_4CH{=}CH_2]_2Br_2$ etc.[312] It was carried out in ethanol on boiling (AIBN initiation). Polymer yield was as high as 42 to 90%. At the same time, attempts to polymerize monomers of type $M(CO)_4[(C_6H_5)_2PH][(C_6H_5)_2PCH{=}CH_2]$ and $M(CO)_4[(C_6H_5)_2PH][(C_6H_5)_2CH_2CH{=}CH_2]$—where M = Cr(VI) or Mo(VI)—in the presence of LiC_6H_5 and free-radical initiators, were unsuccessful.[315] In this case the initiator interacted with the MCM to lead to the deprotonation of the secondary phosphine followed by internal cyclization of the phosphido complex and production of the cyclic compound:

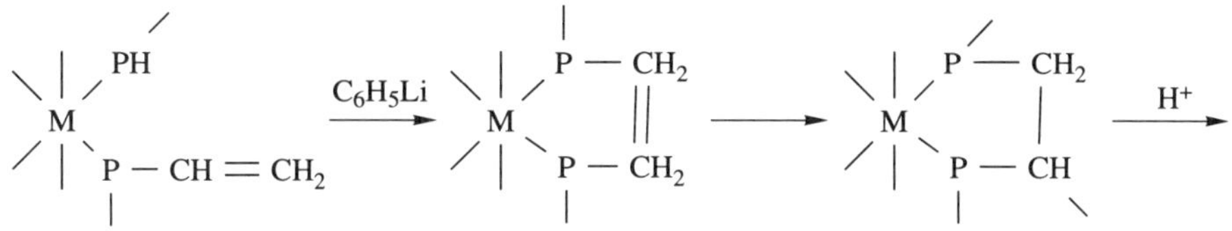

$M(CO)_4[P(C_6H_5)_2CH_2CH_2P(C_6H_5)_2]_2$

Among the problems characteristic of donor-acceptor-type MCM polymerization conversions, a special place is occupied by the conception of complex-radical polymerization.[331] It is based on the assumption of a transition state being formed in the presence of MX_n, where both the propagating radical and the monomer (e.g., MMA) molecules, are coordinated at the same metal atom:

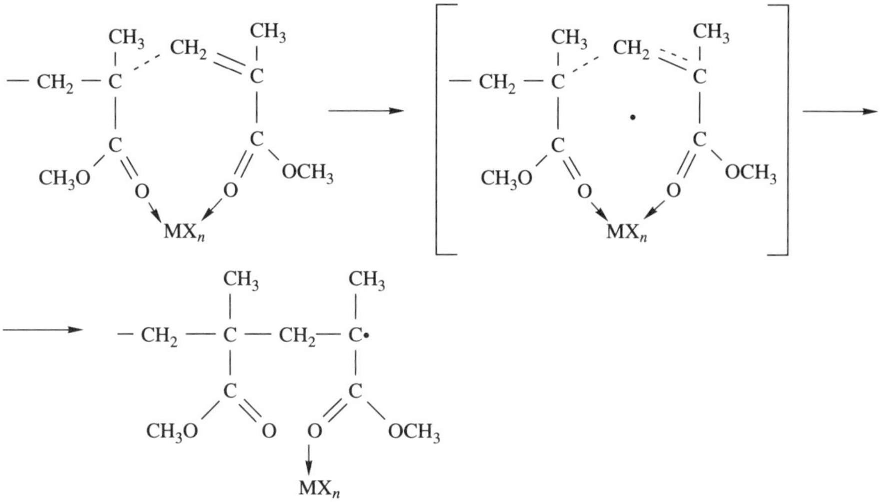

Formation of such a state involves special steric conditions which favor monomer addition to the radical with the result that the polymerization process takes place in the coordination sphere of one metal atom. Along with the steric factor, a large role is played by an energetic contribution in that as a result of a significant delocalization of the lone electron in the intermediate complex E_a decreases. All this leads to increasing k_p. Due to the fact that one MX_n molecule, in principle, can participate in a large number of consecutive propagation acts by migrating to the chain end (the migration being driven by strengthening the "propagation radical-MX_n" bond as compared with the "monomer(M)-MX_n" one), a substantial factor for such a mechanism might be the requirement that the $[MX_n]:[M]$ ratio should be less than unity (optimally, 0.2 to 0.4). It seems likely that for MCM polymerization to take place, this requirement is realizable in the case of strongly dissociating systems only. Another possibility is in the case of a hypothetic mechanism, namely, dissociation in the transition state with a propagating chain. Otherwise, there arise additional hindrances caused by two MX_n molecules taking part in the propagation act. Moreover, it would be hard to obtain a cyclic intermediate state for monomers comprising linear molecules, e.g., in the case of AN:

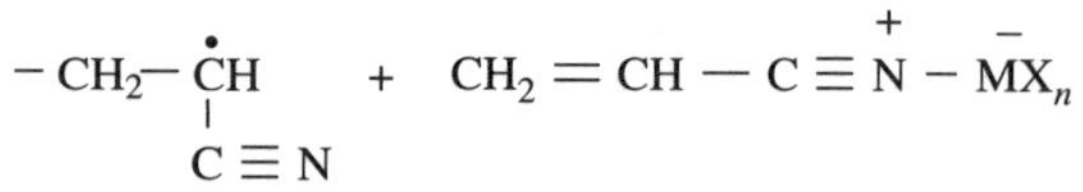

Steric hindrances are rather pronounced for VP and vinylazoles.

There are many works dealing with the complex-radical polymerization of *nV*-type MCMs containing vinyl monomers. However, the most impressive results have been obtained in the polymerization of complex-bonded allyl monomers. It is common knowledge that polymerization conversions of allyl-type compounds present difficulties and usually lead to the formation of low molecular oligomers. This is due to the degradative polymer chain transfer with the appearance of a low-activity allyl radical $CH_2{=}CH{-}\dot{C}H{-}X$. Addition of complexing agents to the systems with allyl monomers affects at least two factors: first, strengthening the C—H bond in the monomer methylene group (with a decrease in the monomer degradative chain transfer rate); and second, enhancing the probability of attaching the resulting complex-bonded allyl radical to the double monomer bond (i.e., to the conversion of the degradative chain transfer into the effective one):

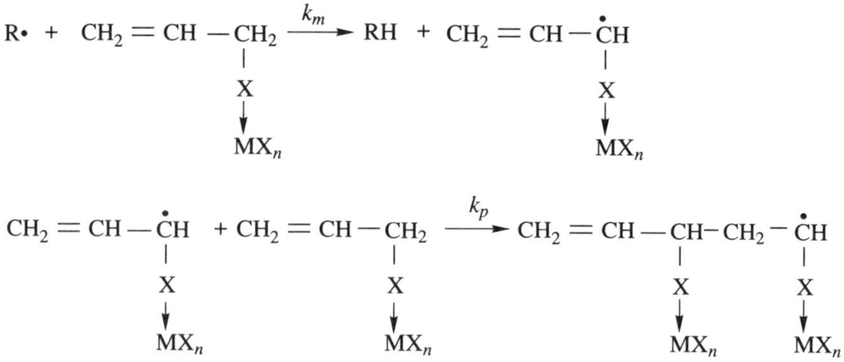

Numerous examples of these effects are known. In particular, as $ZnCl_2$ is added, one can observe an accelerated radiation polymerization of allyl acetate and allylamine (AA) by 6 and 30 times, respectively, with the formation of high molecular polymers.[331] The rate of the $CrCl_3$–AA complex polymerization—75°C, $(NH_4)_2S_2O_8$ initiation—exceeds that of the monomer alone by a factor of 14.8.[332] Decreasing the reaction order with respect to the initiator and the monomer (from 0.5 to 0.4 and 1.2 to 0.8, respectively) is also regarded as one of the indications for changing the degradative chain transfer to the effective one.

B. BULK POLYMERIZATION

In some cases bulk polymerization is advantageous over polymerization in solutions. The absence of solvents is bound to cause high polymerization rates and large MMs of the resulting polymers. This method is especially efficient in the case of monomers with no suitable solvents for conducting the process. Another positive aspect is the possibility of shaping the polymeric product at the synthesis stage. At the same time, block polymerization can be complicated by diffusion hindrances and heat removal problems to eventually lead to various nonisothermal effects.

Among such processes let us mention, first of all, polymerization conversions of 1:1 and 1:2 $ZnCl_2$–MMA complexes which in conventional conditions comprise viscous liquids.[293, 333] These MCMs can be readily polymerized in bulk (AIBN-, γ-, or ultraviolet [UV] initiation). Thus, in the previously mentioned case it was possible to produce polymers with MM = 1 to 2·10^6 and a conversion of up to 85%; whereas γ-initiation gave smaller MMs: M_n did not exceed 39,000 at as high as 96% conversion. The $SnCl_4$-MMA complexes proved less reactive as compared with the $ZnCl_2$-based adducts. Generally, the activity of these complexes in block polymerization conditions changes as follows: $Zn(MMA)Cl_2 > Zn(MMA)_2Cl_2 > Sn(MMA)_2Cl_4 > Sn(MMA)Cl_4 >$ MMA. Also, it should be noted that these MCMs are capable of spontaneous polymerization in the absence of oxygen and radical inhibitors.[333]

In addition, polymerization of some *nV*-type MCMs in melt can be assigned to block polymerization processes. Thus, for a number of complexes the DTA method was useful in detecting a powerful exoeffect suggesting the occurrence of polymerization in these conditions. For example, Co(1-vinyl-2-alkylbenzimidazole)$_2$Cl$_2$ complexes polymerize at 270 to 375°C.[254] In thermal polymerization the activity of these MCMs depends on molecule structure substituents in position (2) of the imidazole cycle: it decreases with the alkyl substituent cycle, and ΔH of the reaction grows from 56.7 to 94.5 kJ/mol depending on substituent nature:

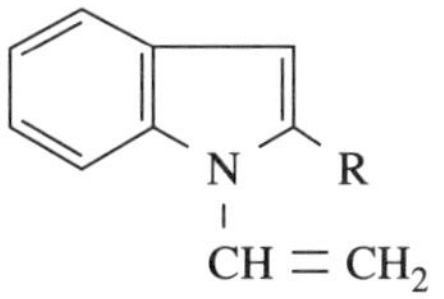

Similar results were obtained for $ZnCl_2$ complexes with the same ligands, the exothermic effect being observed at 245 to 360°C[257] and 280 to 310°C in the case of $CoCl_2$ adducts with VIA and its alkyl derivatives.[247]

C. SOLID-PHASE POLYMERIZATION

The features of solid-phase polymerization as compared with its liquid-phase counterpart were discussed in Chapter 2. Here we shall note only that the intermolecular distance in the monomer crystal is smaller for the uncoordinated configuration. Therefore, for the lattice site vibration amplitude to be sufficient to initiate the polymerization process, the MCM requires higher energy consumption. This leads to larger E_a values for the solid-phase polymerization of *nV*-type MCMs. The values shown below indicate that for the radiation polymerization of *N*-substituted AAm complexes with $ZnCl_2$ and $ZnBr_2$, the E_a value is much higher than that for the individual ligand.[308] For the thermal polymerization of the Zn (*N-tert*-butyl acrylamide[BAAm])$_2$Cl$_2$ complexes it is still larger, 314 kJ/mol.[307]

A comparison of the various initiation methods such as radiation and thermal ones for the solid-phase polymerization of complexes where L are the *N-tert*-alkyl-substituted acrylamides (BAAm, *N-tert*-amyl acrylamide [AAAm], or [HAAm]; X = Cl^- or Br^-[307,308,334] indicates that the principles for these processes are very much the same. Thus in the two cases, the polymerization rate grows significantly with temperature (especially near the melting point [T_m] of the complexes) and each temperature is matched with a limiting conversion. Irrespective of initiation type, the polymerization rates for the $ZnCl_2$ complexes are higher than those for the $ZnBr_2$ ones, which is attributed to steric hindrances because of largely differing anion sizes. Moreover, in most cases (except perhaps for AAAm) in radiation polymerization $ZnBr_2$, as distinct from $ZnCl_2$, produces even a decelerating effect on the process. As this takes place, for different types of initiation the effect of the *N-tert*-alkyl substituent on the polymerization rate at the same temperature will be also the same:

$$Zn(BAAm)_2Cl_2 > Zn(AAAm)_2Cl_2 > Zn(HAAm)_2Cl_2$$

Thus, the polymerization rate increases with substituent length in the homological series of the *N-tert*-alkyl substituents. However, such a correlation is not correct since the complexes have different T_m's the proximity to which is a measure of reactivity in the solid-phase polymerization. In fact, when the reaction takes place at T_m = constant, the role of the substituents will be different. It is obvious that the greater MCM compactness with the shorter alkyl substituent chain length makes the steric hindrances fall in initiated polymerization. Note that the E_a values of the process grow with substituent chain length. It is interesting that the $Zn(BAAm)_2Cl_2$ complex is capable of postradiation polymerization at a rate exceeding that for the uncoordinated ligand. As this takes place, the E_a value of postradiation polymerization is lower than that of the thermal process by ca. 105 kJ/mol.

Polymers of Zn(II) complexes with AAm derivatives—obtained by radiation-induced polymerization in the solid phase—are well soluble in acetone, toluene, and chloroform, which suggests their linear structure despite the presence of two polymerizable ligand groups in the MCM molecule.[308]

As shown by the DTA method, thermal solid-phase polymerization also takes place in the case of $CdCl_2$ complexes with BAAm and AAAm (1:2);[335] however, this process has not been studied kinetically.

Moreover, some 4-VP complexes can be copolymerized in the solid phase. In particular, tetrahedral α-$Co(4\text{-}VP)_2Cl_2$ crystals are thermally polymerizable when heated in air or under nitrogen at 103 to 130°C.[238] The reaction starts at the crystal edges and defects and propagates in the form of an irregular front through the crystal to the completion of the process, the appearance of the crystal remaining unchanged. Polymer yield increases significantly as AIBN is added to the system. It is interesting that unlike the α-form, a high-density octahedral β-$Co(4\text{-}VP)_2Cl_2$ configuration does not polymerize in the solid phase irrespective of what kind of initiator is used.

Attempts to polymerize other VP complexes were less fruitful. Thus, $Co(4\text{-}VP)_4Cl_2$, $Co(4\text{-}VP)_4(SCN)_2$, $Co(2\text{-}VP)_2Cl_2$, $Ni(4\text{-}VP)_4Cl_2$, $Cu(2\text{-}VP)_2Cl_2$, and $Cu(2\text{-}VP)Cl_2$ were found to decompose with heating; whereas polymers that were still obtained with their participation in low yields are closer to copolymers composed of complexes with a smaller CN of metals and of the released VP. At the same time, $Ni(4\text{-}VP)_2Cl_2$ was found to polymerize on the faces of its crystals. The $Cu(4\text{-}VP)_2Cl_2$ and $Zn(4\text{-}VP)_2Cl_2$ complexes polymerize when heated for a long time only at 115 to 120°C. 2-VP complexes proved absolutely nonpolymerizable. The products resulting from the polymerization of the *nV*-type MCMs based on 4-VP have a linear structure.

Solid-phase radiation polymerization of the $Zn(AN)Cl_2$ and $Zn(AN)_2Cl_2$ complexes was effected between –78 and +30°C.[326] For the 1:1 complex the reaction rate was always higher than that for the 1:2 one. The number-average molecular mass (M_n) values for the $Zn(AN)Cl_2$-based polymer was as high as $1.12{\cdot}10^6$. Mention should be made of very low E_a values: 1.1 and 2.2 kJ/mol for $Zn(AN)Cl_2$ and $Zn(AN)_2Cl_2$, respectively. Note that the nature of the propagating chain termination for the two complexes was close to that of the monomolecular reaction ($w \sim [I]^{0.88}$ and $w \sim [I]^{0.99}$ for the 1:1 and 1:2 complexes, respectively). In similar conditions AN-$ZnBr_2$ complexes are also capable of giving macromolecules.

Radiation initiation was also applied to the solid-phase polymerization of the $Zn(MAN)_2Cl_2$ complex at 80°C, the resulting polymer having M_w = 12,000.[277] Participation of another metal complex, namely, $Sn(MAN)_2Cl_4$, was ineffective.

The $Zn(MMA)Cl_2$, $Zn(MMA)_2Cl_2$, $Sn(MMA)Cl_2$, and $Sn(MMA)_2Cl_2$ adducts were polymerized under radiation initiation between –18 and –196°C to obtain products with small MM (M_n = 315 to 5600).[333] For the Sn(IV) complexes polymer conversion was as low as 4 to 5%, and for the Zn(II) species it was higher (15 to 42%). The activities of such MCMs correspond to those in block polymerization.

Postradiation polymerization of transition metal AAm complexes such as $Co(AAm)_4Cl_2$, $Co(AAm)_4(H_2O)_2(NO_3)_2$, and $Ni(AAm)_4Cl_2$ was effected under alcohol matrix devitrification conditions.[156] Polymer yield was as high as 45% but the M_n value of the PAAm isolated from the AAm complex of Co(II) nitrate did not exceed 5600 (in similar conditions this value for the AAm polymerization product is 8700).

D. FRONT POLYMERIZATION

Studies of $M(AAm)_4(H_2O)_2(NO_3)_2$ complexes—where M = Mn(II), Co(II), Ni(II), or Zn(II)—indicate that these monomers are readily polymerizable in the front conditions without a need for adding any initiators or activators.[336] This phenomenon is the first published case of thermally induced front polymerization. As this takes place, the reaction occurs in the mildest possible conditions that are known for such processes. Shown in Figure 21 are dependences of the front propagation rate in $Co(AAm)_4(H_2O)_2(NO_3)_2$ polymerization on the reaction time length from the instant of "starting" at various temperatures. It is evident that there are two zones corresponding to the unsteady and steady modes of front propagation. Depending on the starting temperature, the process rate in the presteady conditions can be either higher or lower than that of the steady front propagation rate w_{fr}.

Studies of the effect produced by a complexing metal on the reaction behavior indicate that w_{fr} decreases in the following series of metals: Co(II) > Ni(II) > Mn(II) > Zn(II).

Note that in the frontal mode the reaction was shown to take place only when AAm complexes of the above nitrates were involved. Uncoordinated AAm; AAm complexes of Cu(II) and Fe(III) nitrates, and transition metal chlorides, bromides, and perchlorates do not polymerize in the same conditions. Thus, this property characteristic of the AAm complexes of Mn(II), Co(II), Ni(II), and Zn(II) nitrates can be regarded one of a type.

Thermal polymerization studies of these MCMs by the DSC method are indicative of the fact that the process runs at a high rate, and a 100% polymer yield is characterized by opening all the four unsaturated groups:[337,338]

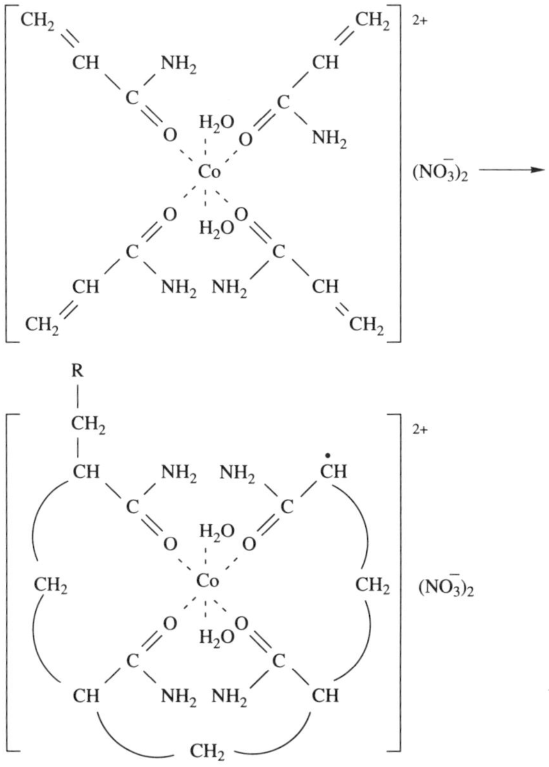

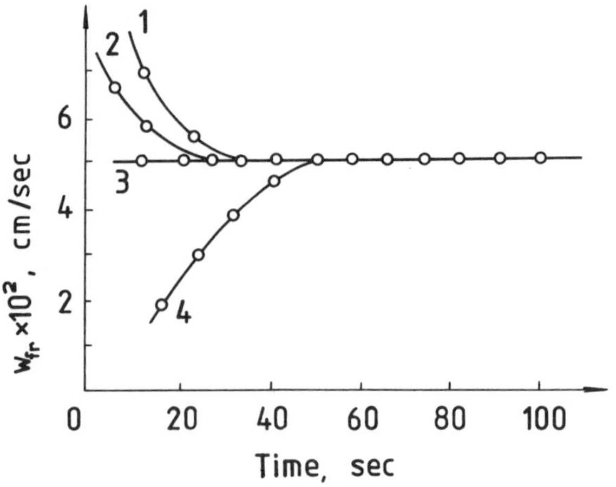

Figure 21 Front travel rate on $Co(AAm)_4(H_2O)_2(NO_3)_2$ polymerization as a function of "starting" temperature and time length of process; d = 4 mm, ρ = 1.02 g/cm³; "starting" temperature, °C: 300 (1), 250 (2), 180 (3), and 120 (4).

The reaction takes place at unexpectedly low temperatures (80 to 100°C) and is distinguished by the fact that the zone of its occurrence lies in the vicinity of the complex T_m (Figure 22, curve 2). At the same time, heating Cu(II) and Fe(III)-based complexes involved oxidation of the organic moiety in the nitrate groups, which is likely to be initiated by decomposition of these groups (Figure 22, curve 3).

In the temperature range over 130°C the Mn(II), Co(II), Ni(II), Zn(II), Cu(II), and Fe(III) nitrates within the monomeric AAm complexes were shown to be thermally unstable. However, those nitrates that are positioned inside the polymeric matrix as a result of polymerization remain stable up to 200°C. In other words, the matrix markedly handicaps their decomposition.

Elementary analysis of the front polymerization products indicates that in terms of composition the resulting polymers are very close to the parent MCM. On removal of the metal nitrate, the M_n value of the PAAm isolated from the polymeric AAm complexes of the Co(II) and Ni(II) nitrates was $9.88 \cdot 10^5$ and $4.14 \cdot 10^5$, respectively; i.e., 100 to 200 times as high as the MM of the polymer obtained by the radical-induced $Co(AAm)_4(H_2O)_2(NO_3)_2$ polymerization in solution.

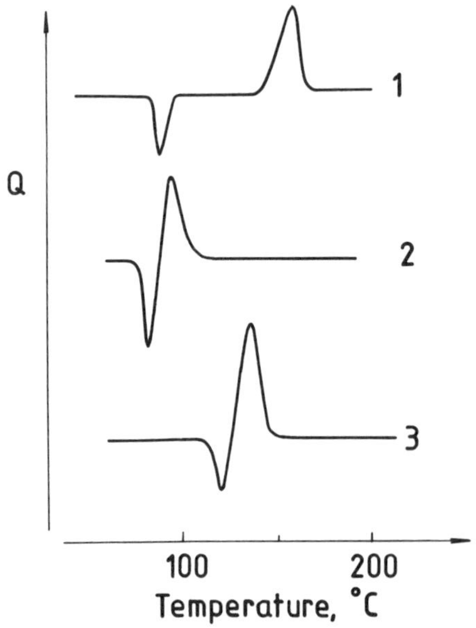

Figure 22 DSC curves for AAm and its complexes: AAm(1), $Co(AAm)_4(H_2O)_2(NO_3)_2$ (2), $Cu(AAm)_4(NO_3)_2$ (3).

Given below is a scheme for initiating the process which incorporates the products of partial nitrates decomposition:

$$Co(II)(AAm)_4(H_2O)_2(NO^-)_2 \rightarrow Co(III)(AAm)_4(H_2O)\left(NO_3^-\right)\ldots\left(NO_2\right)$$

E. POLYMERIZATION OF COMPLEX-BONDED HETEROCYCLES

Along with MX_n complexes including conventional monomers, polymerization conversions are also typical of the case when such complexes incorporate some compounds that generally speaking can hardly be termed monomers. Thus in the last few years, researchers have disclosed the polymerization ability of $ZnCl_2$; and $CoCl_2$, $NiCl_2$, and $PdCl_2$ complexes with ethylene- and triethylene imines and their derivatives (ML_nCl_2)—where $n = 1$ to 4 (preferably, 2).[283-290] Due to the formation of a coordination bond, these complexes are distinguished by rearrangement of the electronic density at the atoms and bonds of the heterocycle with the result that the cycle stress increases appreciably (e.g., for ethylene imine by 42 kJ/mol).[287] An enhanced heterocycle stress in the coordination sphere leads, in turn, to a sharp drop of the endocyclic bond strength. This causes their heterocyclic opening to give bipolar ions with the coordination bonds remaining unchanged. As this takes place, the reaction occurs at relatively low temperatures: 125°C for $ZnCl_2 \cdot 2HNCH_2CH_2$[287] and 116°C for $ZnCl_2 \cdot 2HNCH_2CH_2$[288] (the ligand decomposes at 300 and 350°C, respectively). In the presence of a substituent at the nitrogen atom of the $ZnCl_2 \cdot 2RNCH_2CH_2$ heterocycle, its opening temperature is still lower.[289]

Polymerizations are conducted in block or aprotic solvents at 70 to 130°C.[288, 289] Studies of polymerizable complexes by ESR at elevated temperatures indicate that the reaction mixture contains no free radicals. Moreover, radical reaction inhibitors and initiators do not affect the thermal polymerization of ethylene imine complexes. At the same time, the process is affected by protic solvents leading to the formation of monoaddition products and oligomers.

Polymerization of complex-bonded heterocycles can be described as follows (see the scheme below). It is evident that the presence of at least two heterocycles in the coordination sphere is the key factor for such a process to take place. Complexes with the only heterocycle are capable of giving cyclic dimers alone (piperidine derivatives). Ethylene imine complexes with Ni(II) halides undergo polymerization in two stages.[291] At about 160°C an inner-sphere formation of an $NH_2CH_2CH_2NCH_2CH_2$ dimer is observed. This process is strongly exothermic, the subsequent polymerization at about 180°C being associated with lower heat release.

The products resulting from imine complex polymerizations comprise stable compounds which decompose at temperatures over 250°C (MM = 1,100 to 40,000). In most cases they are well soluble in water and standard organic solvents.

It should be noted that polymerizations of such complexes are only possible when small heterocycles (3- and 4-membered) are involved, for which complexation leads to remarkably strong changes in the electronic density distribution. Polymerization of complexes with larger cycles either does not take place at all (pyrrolidone, piperidine)[290] or, if so, leads to MX_n liberation (pyridine, quinoline).[339]

Such macrocycles as crown ethers behave differently.[340] Although the donor properties of their oxygens do not differ markedly from those of ether oxygen, the macrocyclic

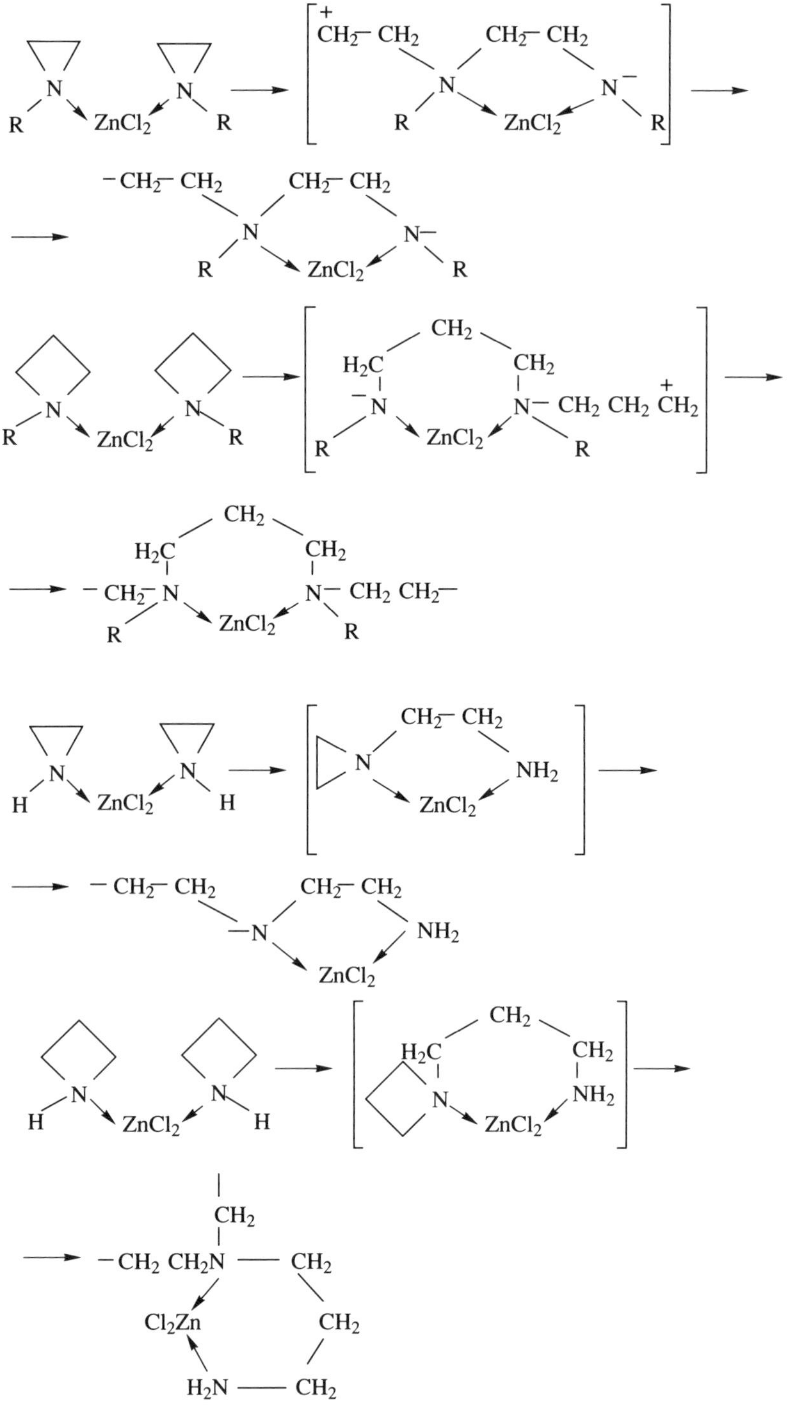

structure distinguished by rigid conformation is responsible for some features in the interaction of the macrocycle with MX_n to result in cycle opening. In this case three products appear at once: (*a*) *nV*-type with the MX_n coordinated via two oxygens; (*b*) these resulting from the crown-ether opening; and (*c*) polymers formed, seemingly, by the cationic polymerization mechanism (the degree of polymerization is 3 to 4):

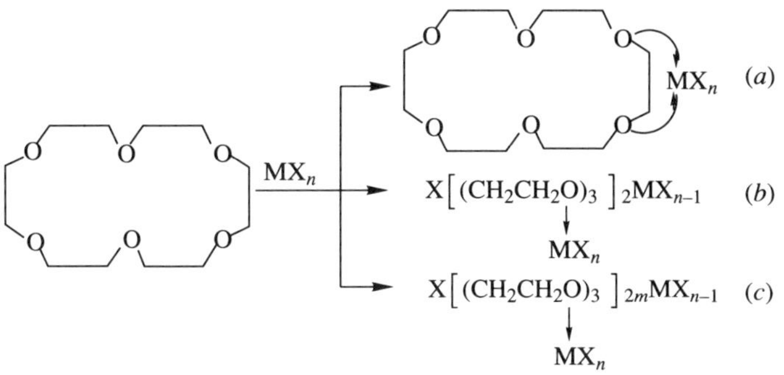

Fractions (*a*), (*b*), and (*c*) in the product resulting from the interaction of dibenzo-18-crown-6 with $TiCl_4$,VCl_4 and $MoCl_5$ are responsible for 87.5, 4.5, 8.0; 60.0, 20.0, 20.0; and 64.0, 24.0, 12.0 wt%, respectively. It is interesting that the interaction of MX_n with the heterocycles readily changing their conformation in the course of complexation does not involve their opening and polymerization. Thus, the interaction of $TiCl_4$ with 1,4-dioxane gives a 1:1 complex with the most probable "chair" conformation.

III. COPOLYMERIZATIONS OF DONOR-ACCEPTOR-TYPE MCMs

In the copolymerization of such MCMs two different cases are likely to take place, depending on whether a metal ion is capable of any additional coordination interactions with a comonomer. The first case is typical of comonomers with properties approximating those of the MCM-contained ligand. In principle, this case may involve situations favorable for a complex-radical polymerization. The limiting case of such a process is a copolymerization of the MX_n complex with an excess ligand. However, in most works relying on this approach, the complexes taking part in the process were not identified.

Although the second case seems to be simpler, there is a certain probability of an additional polarization of the complexed comonomer with participation of the π-electron cloud and the corresponding strengthening of the double bond electron-acceptor properties. The formation of such triple complexes results in the appearance of alternating copolymers.

A. COPOLYMERIZATION OF MCMs BASED ON NITROGEN-CONTAINING LIGANDS

The majority of data concerning copolymerizations of *nV*-type MCMs with conventional monomers relate to complexes incorporating a nitrogen-containing ligand. Thus, a fairly detailed consideration has been given to copolymerization of Zn(II) and MVP complexes with styrene.[239] In the copolymerization of such MCMs one can observe a significant strengthening of the polar factor *e* as compared with an uncoordinated MVP. This effect appears due to the electron-acceptor nature of Zn(II), the coordination with which involves an electronic density redistribution at all the ligand bonds including those of the vinyl group. In this case, a remarkable role is played by the nature of the anion, which decreases differently the electronic affinity of Zn(II) in that the values of *e* are more positive as I^- is replaced by Cl^-. However, the value of I/r_1 representing MCM reactivity with respect to the polystyrene radical does not correlate with the increasing polymerization rate of these MCMs. No explanation of this effect has been given so far.

Another complex, $Zn(4\text{-}VP)_2Cl_2$, was also copolymerized with styrene (UV initiation, $\lambda = 365$ nm).[341] The process was shown to give rise to a charge-transfer MCM-comonomer complex which absorbs much light and generates copolymerization-initiating radicals. The relative reactivity parameters of the process were as follows: $r_1 = 0.24$ and r_2 (MCM) = 2.047. With an increasing dielectric constant of the environment, the rate of the process was found to be lower.

Also, a $(C_4H_9)_3Pd(4\text{-}VP)_2Cl_2$ complex has been polymerized with styrene[319] without giving any copolymerization parameters.

Among the copolymerizations of vinylazole complexes, mention should be made of a successful process with $ZnCl_2$–VBI (VBT) adducts (styrene as the comonomer) (Table 14).[320] It is evident that the activity of the VBI and VBT complexes in reactions with the styrene macroradical ($1/r_1$) is lower than that for an uncoordinated ligand. As this takes place, the order in which the monomer activity changed in the reaction with this radical in complexation with $ZnCl_2$ was opposite to that in which it changed in the case of homopolymerization. This is consistent with classical ideas of radical polymerizations.

Among other vinylazole-based MCMs it would be interesting to pay attention to $M(VIA)_4(OCOCH_3)_2$,where M = Co(II), Ni(II), or Zn(II). Ternary copolymerization of these complexes with *N*-vinylpyrolidone (VPr) and divinylbenzene was conducted in an aqueous buffer solution under radiation initiation.[342] Cu(II) and Fe(III) complexes were found not to enter into copolymerization. An interesting route was proposed for synthesizing actinide-containing copolymers. Such complexes as $UO_2(NO_3)_2$-VIA and $UO_2(NO_3)_2$–[VMIA] were subjected to electrochemical polymerization with excess ligands.[343] Thus, it was possible to obtain products containing over 10% of UO_2(II) with the yield not exceeding 2.2%.

Earlier mention was made of the possible changes in MCM composition during homopolymerizations. The products of such a process can be regarded as off-beat MCM copolymers with a smaller number of coordinated ligands and with a free ligand. This can be exemplified by polymerization of methylvinyltetrazole complexes in ethanol:[265]

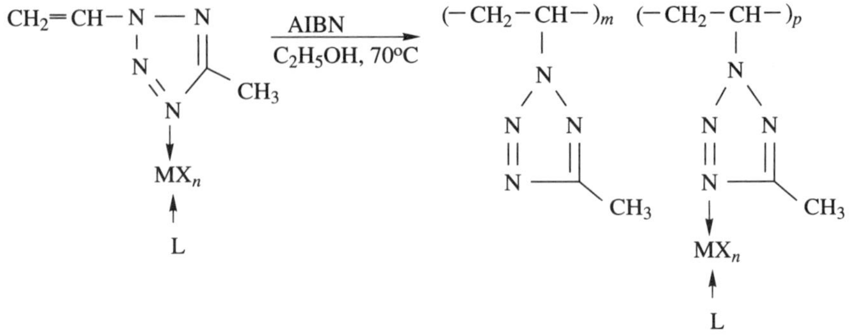

In this situation a "metal-free" analogue is a more active comonomer than the corresponding MCM.

Another example of such a process is a thermal polymerization of complexes with a partial ligand dissociation. This process is used for the polymerization of $Co(4\text{-}VP)_4(SCN)_2$ which decomposes on heating:[242]

$$Co(4-VP)_4(SCN)_2 \rightarrow Co(4-VP)_2(SCN)_2 + 2(4-VP)$$
$$Co(4-VP)_2(SCN)_2 \rightarrow Co(4-VP)(SCN)_2 + 4-VP$$

MCM starts decomposing and the abstracted 4-VP polymerizing at 50 and 104°C, respectively, whereas the maximum polymerization rate is observed at 170°C or so when the $Co(4\text{-}VP)_n(SCN)_2$ (n = 1 or 2) complex remaining after decomposition undergoes

Table 14 **Relative reactivity constants of vinylazoles and their $ZnCl_2$ complexes (M_2) in copolymerization with styrene**[320]

M_2	r_1	r_2	$1/r_1$
VBT	2.5	0.20	0.4
$Zn(VBT)_2Cl_2$	4.6	0.27	0.22
VBI	2.82	0.36	0.35
$Zn(VBI)_2Cl_2$	4.0	0.24	0.25

Note: Copolymerization conditions: DMPA, 60°C, AIBN.

polymerization in the melt. The final product comprises an unidentified molten amorphous blue-colored species. The macroradical from the released 4-VP seems to be capable of initiating the polymerization of the complex resulting from the parent MCM decomposition. The expected result is formation of Co(II)-containing block copolymers.

There are reports on the copolymerization of MCMs containing unsaturated nitriles. Thus, the $Zn(AN)Cl_2$ and $Zn(AN)_2Cl_2$ complexes (benzene, 0 to 30°C, no initiation), as well as the $Sn(MAN)Cl_4$ complex (toluene, –75°C, UV initiation), were found to copolymerize with styrene; the resulting alternating copolymers contained ca. 50 mol% AN (MAN), irrespective of the reaction mixture composition.[269,294] It is assumed that in this case ternary complexes of the type MX_n-AN-styrene are formed. Also, $Zn(AN)Cl_2$ gives alternating copolymers with one more comonomer (VA).[278] The reaction occurs spontaneously at 25 to 35°C in block and comes to completion within ca. 2 h, the process being inhibited by oxygen. At temperatures equal to or in excess of 40°C the product is enriched with AN.

B. COPOLYMERIZATION OF MCMs BASED ON OXYGEN-CONTAINING LIGANDS

Copolymerization of MMA complexes with styrene gives, as a rule, alternating copolymers.[269,294] Thus, in the copolymerization of $Zn(MMA)Cl_2$ or $(Zn(MMA)_2Cl_2)$-styrene systems (benzene, 0 to 30°C, no initiation) the ratio of the initial components in the resulting product was 1:1 or 1:2. A detailed consideration has been given to the copolymerization of $Zn(MMA)Cl_4$ and $Sn(MMA)_2Cl_4$ with styrene (toluene, –75°C, UV initiation), which as opposed to similar $ZnCl_2$-containing complexes causes no cationic styrene polymerization.[344] An NMR method indicates that the reaction mixture contains ca. 30% of labile ternary *nV*-complexes with two configurations:

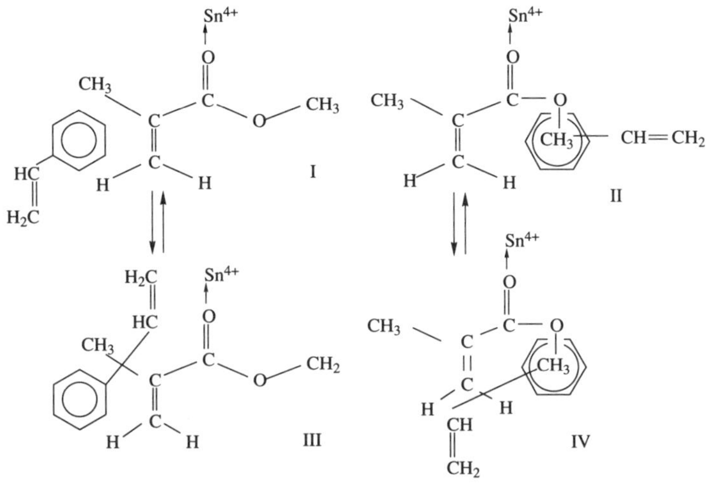

It is assumed that the rotational motion of the styrene molecules can cause a conversion of structures I and II to III and IV, respectively, where the double MMA and styrene bonds crowd together. This fact plays the key role in the ability of such complexes for alternating copolymerization.

It should be noted that the formation of triple complexes is typical of not all MX_n's. Thus, $Sn(MMA)_2Cl_4$ shows the presence of labile coordination bonds. At the stage of chain propagation Sn(IV) interacts with both the π-electronic styrene system and carbonyl MMA groups:[344]

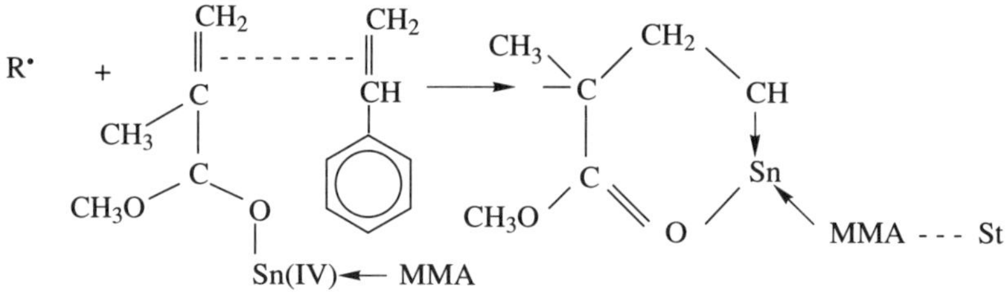

In the case of stronger Lewis acids, e.g., $AlCl_3$, no interaction with the π-electronic styrene system takes place because of their ability to become coordinated with one monomer only. However, in the case of the $Al(MMA)Cl_3$ complex, alternating copolymers with styrene (in mixture with a significant amount of homopolymer) were obtained. At the same time, the $Ti(MMA)Cl_4$ complex gives only traces of copolymer (UV initiation).[294] It is apparent that this leads to the reduction of Ti(IV) with the abstraction of the Cl atom inhibiting the process.

A situation with the production of a copolymer composed of an MMA complex with a comonomer, a ligand of the same complex, can be exemplified by copolymerization of MMA containing a predissolved MX_n (namely, Cu(II) chloride and nitrate, Mn(II) and Pr(III) chloride, Bi(III) bromide and UO_2(II) nitrate).[345] In the case of UO_2(II) nitrate MX_n was responsible for as high as 45 wt%; the remaining MX_n's were less soluble, 3 to 4 wt%. Polymerization was effected at 65 to 70°C in block for 48 h (AIBN initiation). The fact that the products, clear homogeneous glassy polymers, contain two types of carbonyl groups (coordinated and uncoordinated) is corroborated by IR spectroscopy.

Finally, mention should be made of the ability of AAm complexes with Cr(III), Er(III), and other metal nitrates to involve other comonomers such as AA, methacrylic acid (MAA), and maleic anhydride into spontaneous copolymerization in aqueous concentrated solutions.[329] The reaction takes place in the presence of air at room temperature.

C. COPOLYMERIZATION OF MCMs BASED ON SULFUR- AND PHOSPHORUS-CONTAINING LIGANDS

There are reports on MCM copolymerizations with alkyl sulfoxide Cr(III) complexes, namely, $[Cr(R)(OSCH{=}CH_2)_3Cl_3]$ (R = C_4H_9 or C_8H_{17}), without describing the kinetic features of this process.[310]

The phosphorus-containing MCMs of this type are copolymerized by the standard methods in solutions (AIBN initiation). Thus, $(C_6H_5)_2P(C_6H_4CH{=}CH_2)_2CoBr_2$ copolymerizes with styrene in toluene to give a soluble polymer.[312] Another complex, *cis*-$[Pt(C_6H_5)_2PC_6H_4CH{=}CH_2)_2]Cl_2$, gives cross-linked tercopolymers.[313] Copolymerization with styrene and divinylbenzene was conducted in chloroform with an MCM content of 4 to 8% in the mixture (styrene 66 to 94%, divinylbenzene 2 to 30%). When the system contains 2% of the cross-linking agent, the resulting product is still soluble. The NMR method indicates that the polymeric complex retains *cis*-configuration and a certain nonequivalency of the phosphorus-containing groups as in the parent MCM.

IV. ANIONIC-COORDINATION COPOLYMERIZATION OF MCMs WITH OLEFINS AND DIENES

One of the major problems in the chemistry and technology of polyolefins and polydienes is the insertion of various functional groups into their macromolecules with the result that the modified polymers acquire a higher adhesion ability, susceptibility to paints, antielectrostatic properties, etc. In this context, it would be especially interesting to study the copolymerization of olefins with complex-bonded monomers containing various heteroatoms.

A. COPOLYMERIZATION OF ETHYLENE WITH COMPLEX-BONDED VP

A direct insertion of 2-VP, 4-VP, or MVP into catalytic systems such as MX_n-AlR_xCl_{3-x} (MX_n = $TiCl_4$, VCL_4, $VOCl_3$, etc.) for ethylene polymerization leads to a complete inhibition of this process.[240] This is attributed to a strong VP interaction with the active polymerization sites.

Chain propagation in anionic-coordination polymerization is a two-stage process to include monomer coordination at the active site "vacancies" followed by insertion of the activated monomer at the M—C bond. For the heteroatom-containing monomers their insertion into the propagating chain leads to the active site suppression and ethylene polymerization interruption by the "coordination-insertion" mechanism:

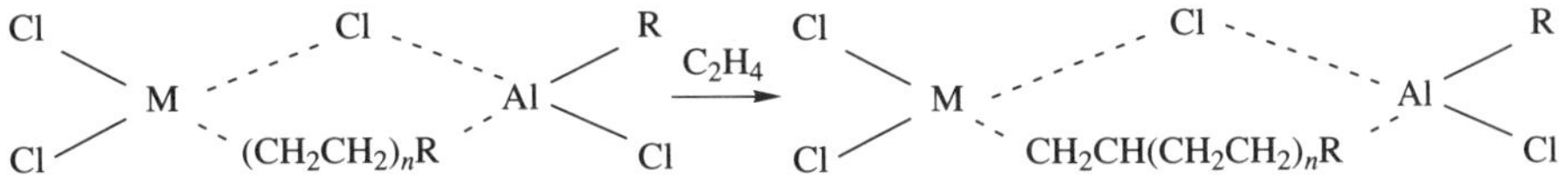

However, the ability of such monomers for coordination can be relied on to involve them into anionic-coordination copolymerization with ethylene. To do so, catalytic system components such as transition metal halides should be used as MX_n's. Such an approach enables one to neutralize the inhibiting action of the heteroatom on the catalyst and to activate the exocyclic bond.

In fact, VP complexes with vanadium halides in combination with organoaluminum compounds are very efficient for ethylene polymerization (Figure 23).[240] Their activity

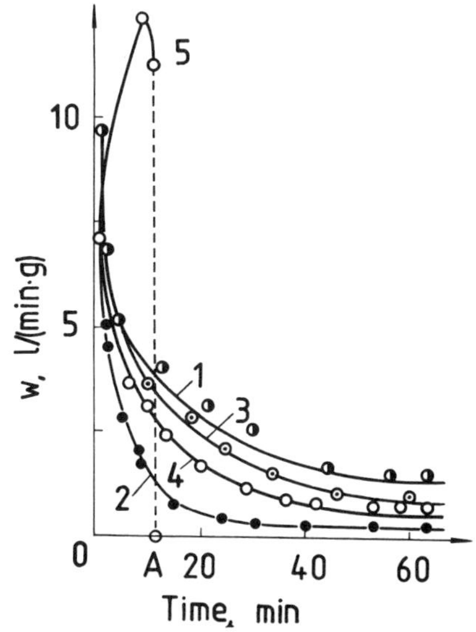

Figure 23 Kinetics of catalytic ethylene copolymerization with vinylpyridines complex-bonded to VCl_4: $V(4\text{-}VP)_2Cl_4$ (1), $V(2\text{-}VP)_2Cl_4$ (2), $V(MVP)_2Cl_4$ (3), $V(4\text{-methyl-5-}VP)_2Cl_4$ (4), termination of ethylene polymerization with VCl_4-Al (*iso*-$C_4H_9)_2$ Cl system on addition of 2-VP (5).

grows with temperature (20 to 60°C), ethylene concentration, and an [Al]:[V] molar ratio (up to 10 to 15°C), as well as with using aromatic hydrocarbons as the solvents. For the VL_2Cl_4 complexes the activity is strongly related to the ligand nature: 4–VP > MVP > 4–Me–2–VP > 2–VP. The Ti $(4\text{–}VP)_2Cl_4$ complex is very strong; therefore, its activity is ca. 200 times lower than that of V $(4VP)_2Cl_4$. Formation of ethylene complexes seems to be encouraged by a higher polarity of the double bond as a result of its electrons being shifted to the heteroring side.

The resulting ethylene copolymers contained 0.1 to 0.9 wt% of nitrogen. In other words, one VP unit is matched with as many as 80 to 120 ethylene ones. Most probably in such copolymers the units are distributed randomly so that the VP fragments are separated by ethylene blocks.

B. CATALYTIC COPOLYMERIZATION OF ETHYLENE WITH NITRILES

Homopolymerization of AN on metallocomplex catalysts is a low-efficiency process in that it gives in insignificant yields a low MM polymer (ca. 2500) with a partially cyclisized structure. Using the approach described above, ethylene can be copolymerized with the corresponding $Ti(AN)_2Cl_4$ and $V(AN)_nCl_4$ complexes ($n = 2$ or 3). However, as opposed to VP complexes, this copolymerization is practicable when use is made of AN complex-bonded to Lewis bases which do not form part of $Al(AN)_2Cl_3$, $Sn(AN)_2Cl_4$, and other catalysts. Analysis of the polymers obtained on the $V(AN)_nCl_4$-Al $(iso\text{-}C_4H_9)_2Cl$—$AlCl_3$ system suggests the presence of 0.5 to 1.2 wt% of nitrogen therein.[270,271,346] The IR spectra of the resulting products indicate that the complex-bonded AN copolymerizes partially due to opening the C=C bond and partially at the C—N group. This is supported by the appearance of absorption bands at 1600 to 1700/cm assigned to the vibrations of the appearing C=N bond. Due to the fact that this process does not involve AN homopolymerization in these conditions, it can be concluded that the AN units are distributed rather evenly and do not form aggregates.

It should be noted that the complex-bonded saturated nitriles do not copolymerize with ethylene, although such a possibility is energetically justified since the C=C and C=N bond energies are close to each other: 273 and 265 kJ/mol, respectively. It might be due to the fact that during complexation the C≡N bond becomes stronger.[346]

C. COPOLYMERIZATION OF COORDINATED MMA WITH OLEFINS

The mechanism of coordinated MMA copolymerization on complex catalysts, as well as other vinyl-type monomers, is not clear. It is not even known whether this process is radical. Generally, the behavior of coordinated MMA does not differ from the polymerization of VP and AN complexes. The insertion of MMA into the reaction volume after formation of a catalyst (e.g., in the VCl_4–AlR_2Cl system even at an [MMA]:[VCl_4] ratio of 1:1) leads to a full inhibition of ethylene polymerization.[347] However, the results of radiochemical analysis indicate that despite the termination of ethylene polymerization, MMA forms part (as before) of the copolymer. In other words, one can observe the emergence of a polyethylene(PE)-grafted (PPMA). The mechanism of this process constitutes a free-radical polymerization due to radical generation in the homolytic splitting of the V—C bond in the presence of MMA:

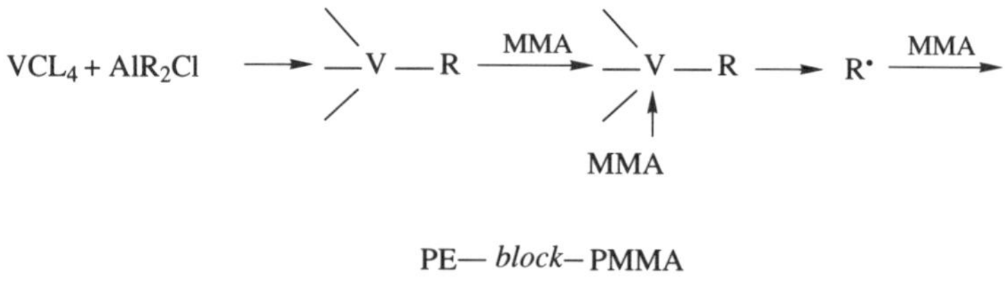

The resulting copolymers contain PMMA blocks composed of 27 to 84 monomer units.

Thus, MMA plays the role of a "switching" agent in the inversion of the anionic-coordination polymerization mechanism to the radical one in the presence of metallocomplex systems. On the other hand, in the case of the complex-bonded MMA—the $V(MMA)Cl_4$—AlR_2Cl system—no regeneration of the active sites takes place; and copolymerization involves the MMA complex-bonded to AlR_2Cl as a result of internal recoordination processes.[292]

Note that the modification of the $TiCl_3$ surface with MA leads to the formation of $Ti(MA)Cl_3$ complexes on the catalyst surface, which exhibit activity in stereospecific propylene polymerization.[348] The associated copolymerization of the complex-bonded MA with PE gives rise to a product with a heptane-insoluble fraction that contains 89% of MA units. Among other things, this constitutes an experimental indication of the presence of various active sites of propylene polymerization on the catalyst surface.

D. COPOLYMERIZATION OF OXYGEN-CONTAINING MONOMERS, OR PRODUCTS RESULTING FROM THE INTERACTION OF VINYL MONOMERS WITH ORGANOALUMINUM COMPOUNDS

Early reports on the modification of metallocomplex catalysts indicate that formation of active sites in such systems is preceded by an exchange recoordination.[349] For a $V(MMA)Cl_4$–AlR_xCl_{3-x} system quantitative parameters of such processes have been determined.[292] Therefore, it proved suitable to effect a prepolymerization binding of polar vinyl monomers with the second component of the catalytic system, namely, an organoaluminum compound. Such a product would serve simultaneously as a comonomer, a component of the catalyst, and its modifier. This approach was realized, for example, in butadiene copolymerization with a complex-bonded AN on a $VOCl_3$-Al (AN) $(C_2H_5)_2Cl$ system,[350] as well as when using an AlR_2Cl system with alkyl-substituted acrylates[351] and MMA.[352] The copolymerization ability of such MCMs correlates with the electronic density at their double bonds and with the strength of the resulting complexes. Another modification of this approach consisting of olefin copolymerization with Al(III)-containing salts of unsaturated acids was discussed earlier (see Chapter 2).

V. GRAFT COPOLYMERIZATION OF DONOR-ACCEPTOR-TYPE MCMs

A major proportion of data on the graft polymerization of *nV*-type MCMs is concerned with complexes of unidentified composition. Thus, there are reports devoted to grafting in the presence of $ZnCl_2$, $ZnBr_2$, and $AlCl_3$ in the following systems: AN-gelatin ($K_2S_2O_8$ initiation),[353] and AAm, MMA, or VA-styrene-1,2- or 1,4-polybutadiene (AIBN initiation).[354] Metal content in the grafted copolymers was not determined. The complexes of VP with Ni(II) salts at an initial molar ratio of 1:1 to 1:10 when grafted onto synthetic and natural rubbers (radical initiation) in solution at 60 to 100°C were not identified.[355]

One can note a fairly high efficiency of grafting allyl monomers complex-bonded to MX_n. Thus for grafting allyl amine, allyl acetate, allyl alcohol (AAl), and allyl mercaptan; and AA, MAA, AN, and MAN esters coordinated to LiCl, $CaCl_2$, and $ZnCl_2$ (a molar ratio of 1:0.5 to 1:5) to polyolefins, the reaction mixture was exposed to irradiation.[356,357] Postirradiation graft polymerization to PE powder (preirradiation in air) was effected in the case of a $Co(AAl)_2Cl_2$ complex (D = 200 kJ/mol, I = 2.8 J/kg·s, no solvent).[358] At 70°C the process took place at a constant rate within a fairly wide time interval (up to 9 h). As the $CoCl_2$ concentration in the system was increased, the rate grew linearly to $[CoCl_2]:[AAl]$ = 1:5 to reach a limiting value; E_a = 40 kJ/mol.

In conditions similar to those when transition metal acrylates were grafted (see Chapter 2), there are reports on the graft polymerization of some AAm complexes such as $Co(AAm)_4Cl_2$, $Co(AAm)_4(H_2O)_2(NO_3)_2$, $Ni(AAm)_4Cl_2$, $Cu(AAm)_4Cl_2$, $Cr(AAm)_4Cl_3$, and $Mn(Am)_4Cl_2$.[233,359] As in the case of acrylates, the content of the homopolymers from these MCMs in the products was low: one fourth to one eighth as high as with a "metal-free" analogue (i.e., AAm). An exception to this was the Mn(II) complex for which the rate of thermal homopolymerization exceeded that of grafting. However, in this process the activities of MCM based on various metals were arranged differently: Cr(III) > Co(II) > Ni(II) (Figure 24). During the grafting the $CuCl_2$ complex was absolutely inactive. The process was considerably affected by the nature of the anion in that $Co(AAm)_4(H_2O)_2(NO_3)_2$ was polymerized at a rate exceeding that of grafting a similar $CoCl_2$ complex by a factor of about three. The grafting rate was proportional to [MCM] in the solution (for an uncoordinated AAm $w \sim [AAm]^{1.7}$). Graft polymerization of AAm and its complexes involved monomolecular termination of the propagating chains; the E_a values for $Co(AAm)_4Cl_2$ and $Ni(AAm)_4Cl_2$ were 51 and 60 kJ/mol, respectively. As in the case of acrylate grafting, the process with participation of AAm complexes was affected by radical initiation inhibitors.

In conclusion, it is evident that donor-acceptor-type MCMs comprise promising objects for polymerization conversions. As a rule, their synthesis is very simple; and it is not accidental that at present the coordination chemistry of VP, vinylazoles, and AAm is well understood. The complexation of O-, S- and P-containing monomers has been studied less.

As compared with syntheses, polymerization conversions of donor-acceptor-type MCMs have been given much less attention. This is especially true of vinylazole complexes. Note the existence of considerable problems concerning the optimization of polymerization conditions for such MCMs. Thus for reactions in solution, the complex is very likely to dissociate so that the resulting polymer will be short of MX_n as compared with the parent MCM. In this case a positive role can be played by using solid-phase, as well as a number of nontrivial methods of initiation and polymerization: conducting the reactions under high-pressure conditions, in the front mode in melt, etc.

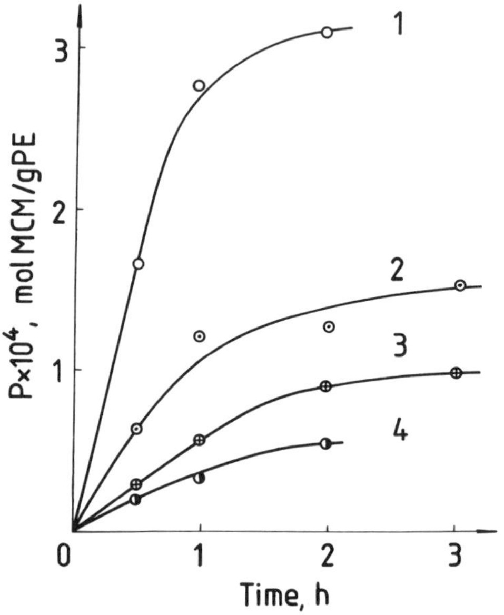

Figure 24 Kinetic curves for yield of polymers resulting from grafting of AAm complexes onto PE: $Cr(AAm)_4Cl_3$ (1), $Co(AAm)_4Cl_2$ (2), $Ni(AAm)_4Cl_2$ (3), $Co(AAm)_4(NO_3)_2(H_2O)_2$ (4).

Generally, it should be noted that polymerization conversions of *nV*-type MCMs (homo-, co-, and graft polymerizations) have received unduly little attention. There is still a question of changing the coordination bond strength in the course of MCM polymerization. Data on changing the reactivity of MCMs as compared with that of an uncoordinated ligand are conflicting.

Chapter 4

Chelate- and Cluster-Type Metal-Containing Monomers

For the time being, chelate- and especially cluster-type metal-containing monomers (MCMs) in terms of their polymerization conversions, are species with potentials that are to be realized in the near future. The synthesis of such MCMs is usually a hard to accomplish process, as compared with their ionic- or *nV*-type counterparts. Another problem is structural identification of the resulting products.

I. METALLOCHELATE-TYPE MONOMERS

These monomers include compounds containing a metal-chelate cycle and an openable multiple bond. According to metal ion-chelating fragment bonding, the MCMs under discussion can be classed substantially as follows:[360]

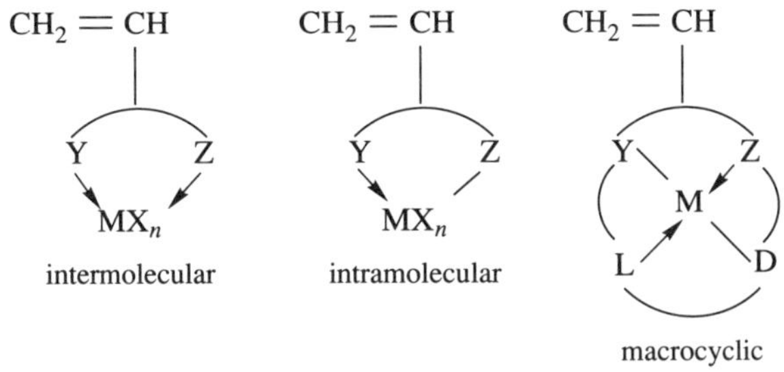

Intermolecular-type chelate MCMs are only characterized by coordination bonds of a metal with the chelating donor atoms. In intramolecular monomeric metallochelates at least one fragment forms both *nV*- and σ-bonds with the metal ion.

A. SYNTHESIS OF CHELATE-TYPE MCMs

A conventional method of synthesizing intermolecular-type metallochelate monomers relies on the metal compound (MX_n) reacting with the chelating monomers. Such a method was used to obtain complexes of Cu(II) chloride with vinyl monoethanolamine ester;[36] Ru(II), Co(II), and Fe(III) with 4-vinyl-, 6-vinyl-, or 4′-vinylterpyridines;[362], Ru(II), Zn(II), and Ir(III) with 4-vinyl-4′-methyl-2,2-dipyridyl;[363-366] Ru(II) with 6-vinyl-2,2-dipyridyl;[366] Cu(II), Ni(II), Co(II), Mn(II), and Zn(II) with 1-vinyl-2-hydroxymethylimidazole;[367,368] Cu(II) with ethyl-α(acetoacetoxymethyl) acrylate,[369] and Ni(II), Co(II), Cd(II), UO_2(II), and Cu(II) with 5-vinylsalicylideneaniline.[370]

Another popular method is based on the interaction of monofunctional monomers with metallochelates. It was instrumental in producing Ru(II) and Os(II) chelate monomers incorporating dipyridyl (Dipy) or 1,10-phenanthroline (Phen)-type ligands as chelating agents, and 4-vinylpyridine(4–VP), *bis*-(4-pyridyl)ethylene, *trans*-stilbazole, *N*-(4-pyridyl)acrylamide, etc. as monofunctional species.[371-375]

One more method for producing metallochelates of this type is based on the reaction of the earlier described MCMs with chelating agents. This can be exemplified by the interaction of Dipy or Phen with transition metal acrylates:[376]

$$M(OCOCH{=}CH_2)_n \cdot mH_2O + L \rightarrow M(OCOCH{=}CH_2)_n \cdot L \cdot pH_2O,$$

where M = Cu(II), Ni(II), Co(II), or Cr(III); m, p = 0 to 2; n = 2 or 3.

Intramolecular chelate-type MCMs can be obtained by a direct interaction of MX_n with monomers containing chelating fragments. Thus, it was possible to generate Cu(II) complexes with *N*-vinylsalicylaldimines,[377] as well as transition[378-380] and alkaline-earth[381] metals with vinyl-β-ketones, palladium with *N*-allylsalicylaldimine,[382] etc. Also, there are more intricate "assembly" methods such as:[383-385]

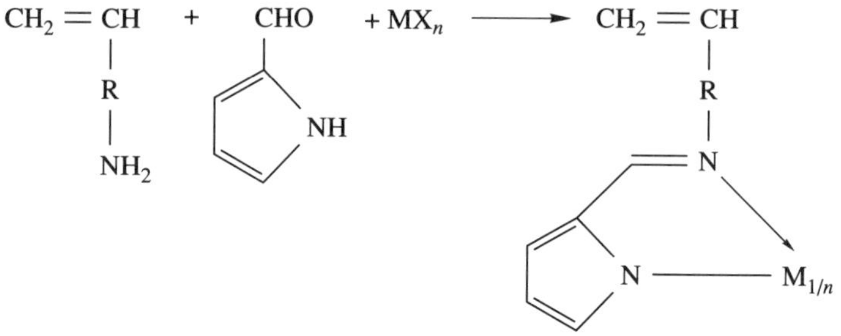

where R = —C_6H_4—, —CH_2—; M = Cu(II) or Co(II); n = 2 or 3. An acid-base reaction was used to synthesize an Eu(III)-containing chelate-type MCM:[386]

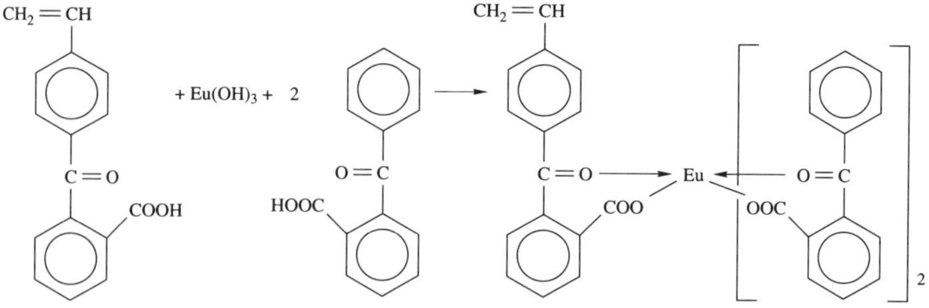

One of the methods for generating macrocyclic metallochelate MCMs consists of a direct interaction of MX_n with macrocycles containing unsaturated bonds. It was employed to obtain vinylporphyrin Co(II), Ni(II), Fe(III), Fe(II), Mg(II), Zn(II), and Cu(II) complexes incorporating one or two side groups:[387, 388]

(*a*) metalloderivatives of protoporphyrin IX

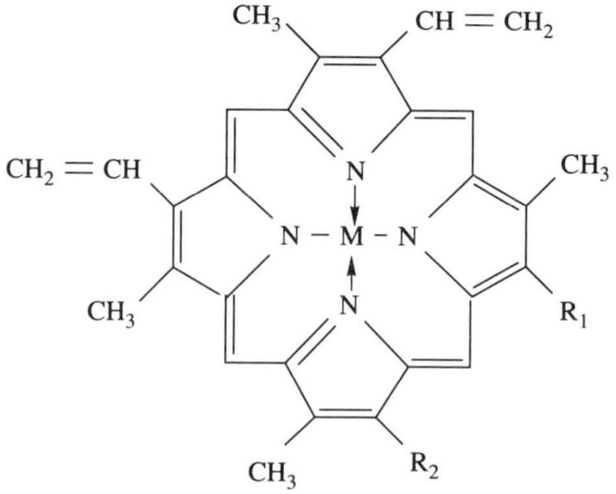

(*b*) metalloderivatives of *p*-vinylbenzyl chlorinates

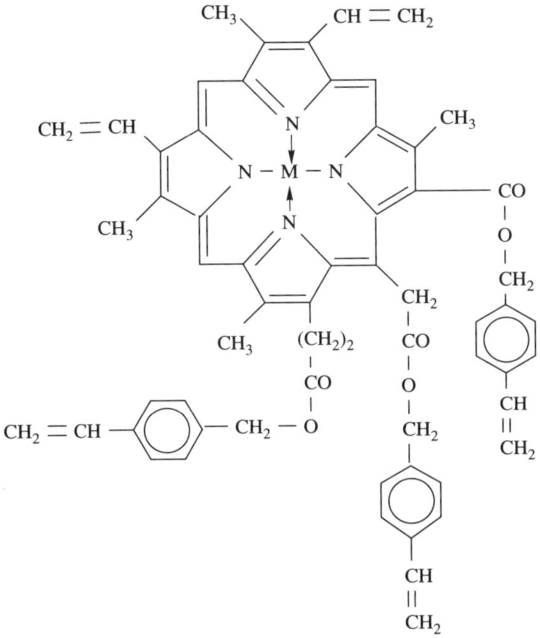

(*c*) metalloderivatives of tetra(*p*-styryl)-porphyrin or acrylamidotetraphenylporphyrin

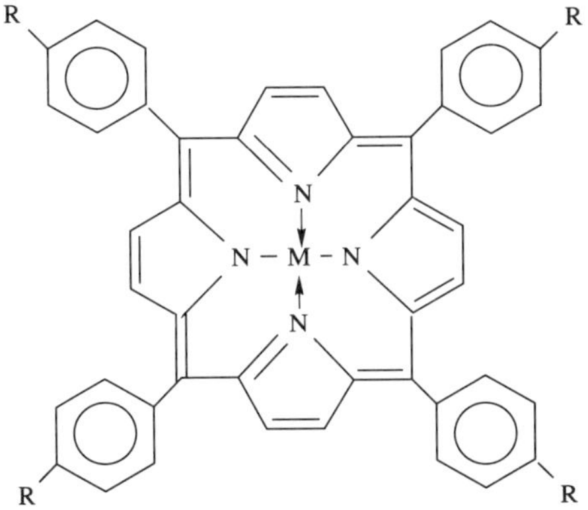

However, metalloporphyrin monomers are largely synthesized through interaction of reactive peripheral groups of transition metal porphyrinates (e.g., chlorophyll, chlorine, and similar metallocomplexes) with conventional monomers:[388,389]

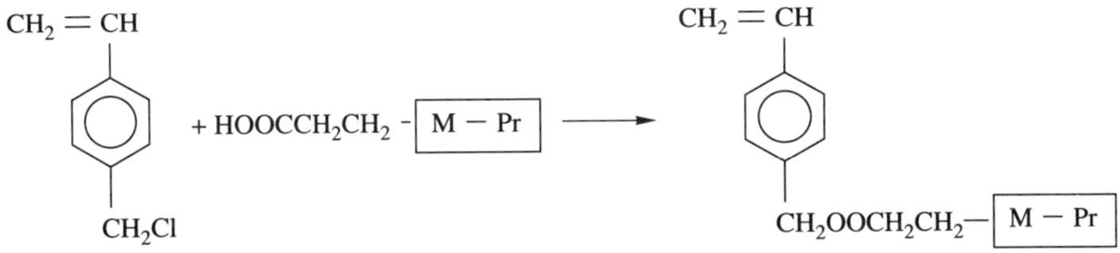

where M—Pr is metalloporphyrin. Carboxyl-, sulfo-, and aminogroups represent the most widely used peripheral species of this kind. Similar monomers can be synthesized by reacting acryloyl with Mn(II) tetra-*p*-aminoporphyrinate.

The above approaches also have been applied to the synthesis of metallophthalocyanine monomers. Thus, there are reports on a Co(II) phthalocyanate synthesis (M—Pht is metal phthalocyanate):[390]

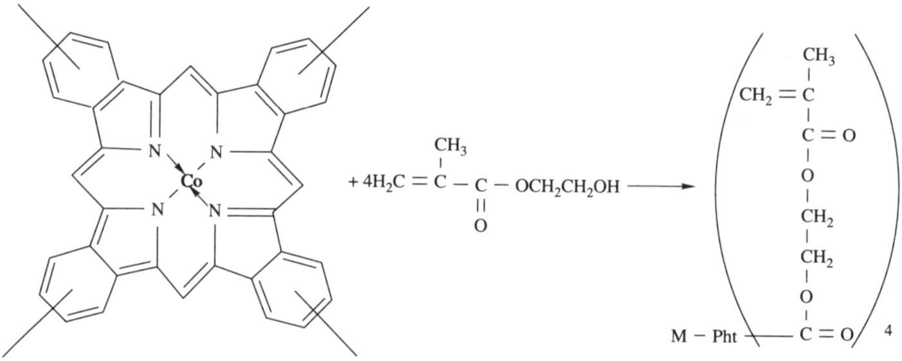

A reaction of methacryloyl chlorides with metal tetra-*p*-carboxyphenyl phthalocyanates is equally useful for this purpose.

B. HOMOPOLYMERIZATION OF CHELATE-TYPE MCMs

Information concerning the polymerization conversions of metallochelate monomers has, in fact, appeared in the last few years. Thus the literature reports on the possibility of polymerizing *cis*-[Ru(Dipy)$_2$(4–VP)X]$^{n+1}$ complexes, where $n = 1$, X = Cl; $n = 2$, X = CO or 4–VP.[371]

Radical polymerization has been applied to the production of Er(III) and VO(II) poly(*p*-acryloxyphenyltriphenylporphyrinates), the elemental analysis corroborating the fact that the metal ions are responsible for 100% of the coordination sites in the product.[391] (Note for comparison that in the case of a similar metal-containing polymer being synthesized by polymer analogous conversions, no more than a 50% metal binding was observed.)

Also, Cu(II), Ni(II), Co(II), Cd(II), and UO_2(II) were polymerized with 5-vinylsalicylidene aniline by boiling in dimethylolpropionic acid (DMPA) in the presence of azobisisobutyronitrile (AIBN).[370] Depending on anion nature (nitrate or acetate), in the case of UO_2(II), the following polymers were obtained:

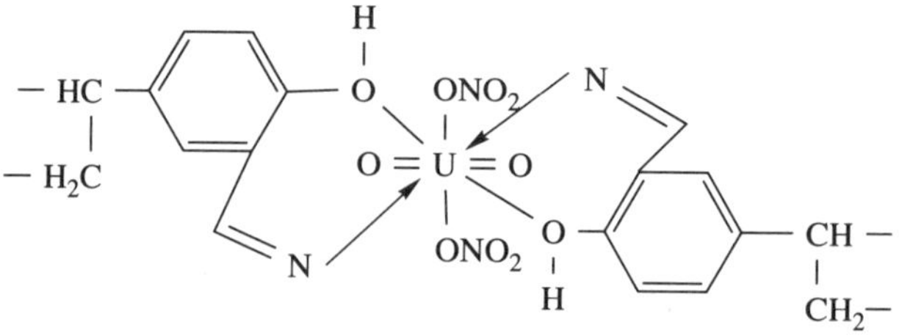

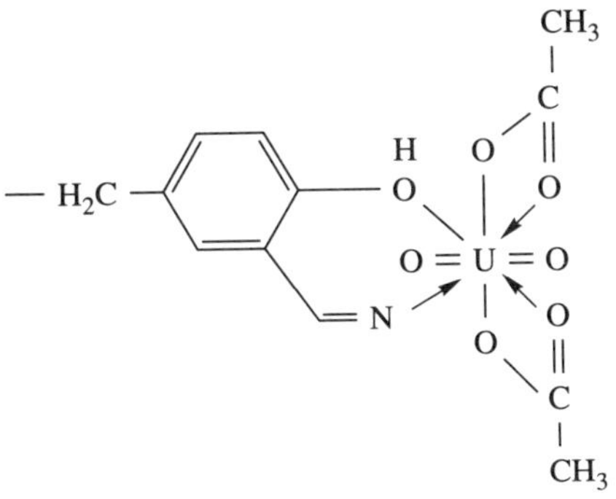

In addition, salicylaldimine Cu(II) chelates:

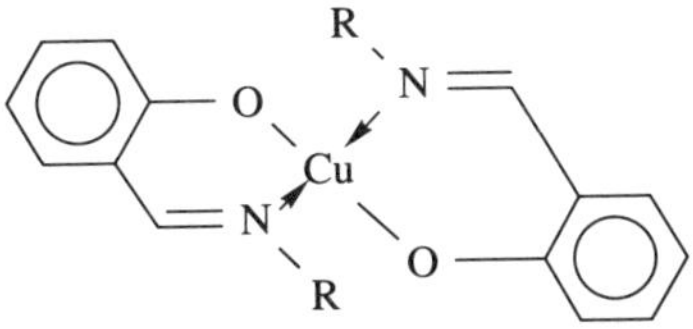

where R = 4—CH_2=CH—C_6H_4—, 4—$CH_2C(CH_3)$—COO—C_6H_4—, and/or CH_2=$C(CH_3)COOCH_2CH_2$— were polymerized at very high initiator concentrations ([AIBN]/[MCM] = 0.12).[377] As this took place, the rate of the process proved far lower than that in the polymerization of various parent salicylaldimines. Another Cu(II)-containing chelate based on ethyl-α(acetoacetoxymethyl)acrylate:

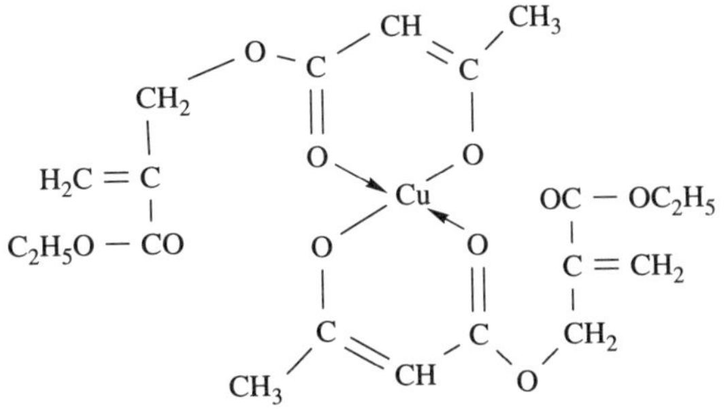

are also hard to polymerize (tetrahydrofuran [THF], 40°C, 2,2′azo-2,4-dimethylvaleronitrile initiation).[369] In 20 h the conversion was no more than 7.8%, and the product had a low number-average molecular mass, M_n = 2100, M_w = 2600.

At the same time, some of the Cu(II)-containing chelate monomers cannot be polymerized at all. Thus, attempts to polymerize Cu(II) methacroyl acetonate in the presence of various initiators were unsuccessful.[378] A low homopolymerization efficiency for such MCMs can be attributed to the intramolecular chain termination as in the case discussed in Chapter 2:

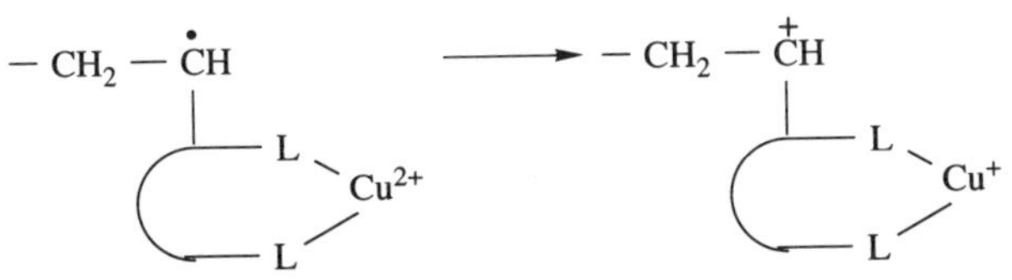

Note that radical polymerization cannot be applied to allyl derivatives of metal β-diketonates either.[380]

A more detailed consideration has been given to the electrochemical reductive polymerization of perchlorate and mixed-ligand hexafluorophosphate chelate monomers of Ru(II), Os(II), and Fe(II) that contain chelating fragments of the Dipy, Phen, and 2, 2′,2′′, tripyridyl (Tpy) types and also a monofunctional monomer (namely, 4–VP and *trans*-4-stilbazole) to name a few.[372, 373, 392] The polymerization was performed by way of repeated cyclic changes in the electron potential around a value corresponding to ligand reduction. As this takes place, the electrode surface gets covered with a uniform polymeric film. A similar process occurs when polymerizing monomers in which the vinyl group is bonded to the chelating fragment, e.g., Zn(II), Ru(II), and Ir(III) complexes with 4-vinyl-4′-methyl-2,2′-dipyridil.[364] In this case, the filming efficiency (the number of polymer moles per mole of parent monomer) is significantly higher.

The electrochemical polymerization mechanism includes formation of the radical ion:

$$\left[\mathrm{Ru(Dipy)_2(L)_2}\right]^{2+} + e^- \rightarrow \left[\mathrm{Ru(Dipy)(Dipy^{\bullet-})(L)_2}\right]^{+} \xrightarrow{+e^-} \left[\mathrm{Ru(Dipy^{\bullet-})_2(L)_2}\right]^{0}$$

The process is initiated by one-electron reduction based on the intramolecular redox equilibrium:

$$\left[\mathrm{Ru(Dipy)(Dipy^{\bullet-})(L)_2}\right]^{+} \rightleftarrows \left[\mathrm{Ru(Dipy)(Dipy)^0(L)(L^{\bullet-})}\right]^{+}$$

In this case, an increase in the number of vinyl groups in the complex leads to a noticeable growth of the polymerization rate.[362]

Note also the possibility of thermal polymerization of the complex made up of 5,5′-bis(3-acryloyl-1-propoxy)carbonyl-2,2′-dipyridyl and Ru(II):[393-395]

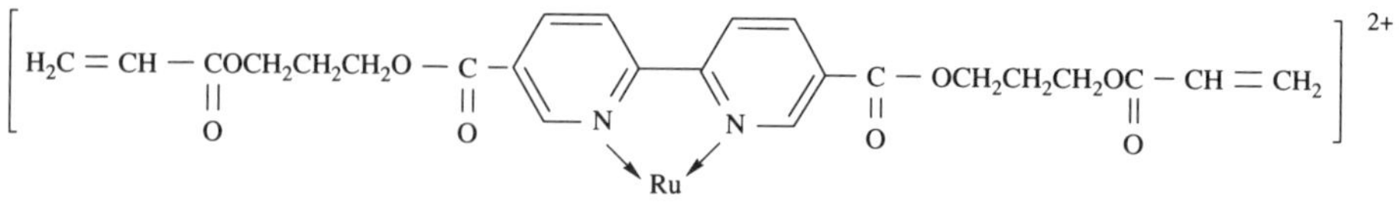

C. COPOLYMERIZATION OF CHELATE-TYPE MCMs

As in the case of the above types of MCMs, copolymerization with conventional monomers represents a more routine practice than their homopolymerization. Thus, methyl methacrylate (MMA) was copolymerized with Eu(III) and Tb(III)-containing monomers in block or in methanol:[381]

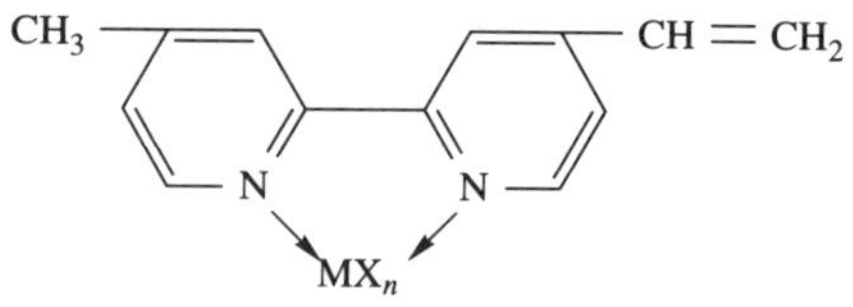

where X = Dipy, Phen, Tpy, etc. Other efforts included styrene, and MMA, as well as 4–VP copolymerized with such complexes as *cis*-$[Ru(Dipy)_2(4\text{–}VP)Cl]ClO_4$ or *cis*-$[Ru(Dipy)_2(4\text{–}VP)_2](ClO_4)_2$.[371] In this case, the copolymers based on Ru(II) complexes containing a 4–VP molecule were soluble in dichloroethane and methanol; whereas the copolymerization product, namely, *cis*-$[Ru(Dipy)_2(4\text{–}VP)_2]ClO_4)_2$ (comonomer 4–VP, the initial ratio [4–VP]:[MCM] = 20), was insensitive to organic solvents since the polymer chains were cross-linked via the metal atoms:

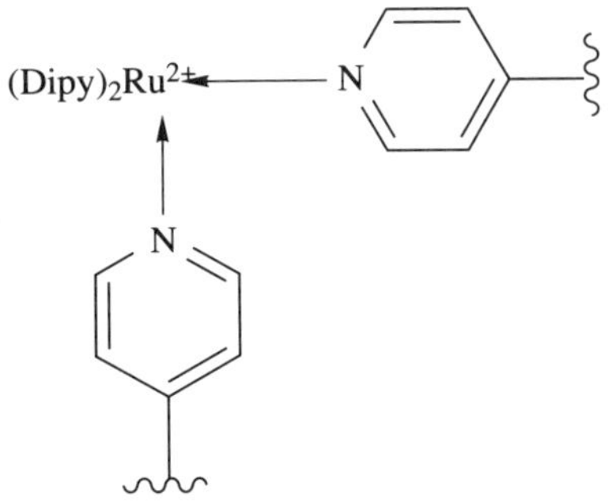

However, such a cross-linking effect of MX_n becomes much weaker as the MCM content decreases.

One more MCM, 2-(4′-vinylbenzoyl benzoate)-di(2-benzoyl benzoate) of Eu(III), was copolymerized with styrene:[386]

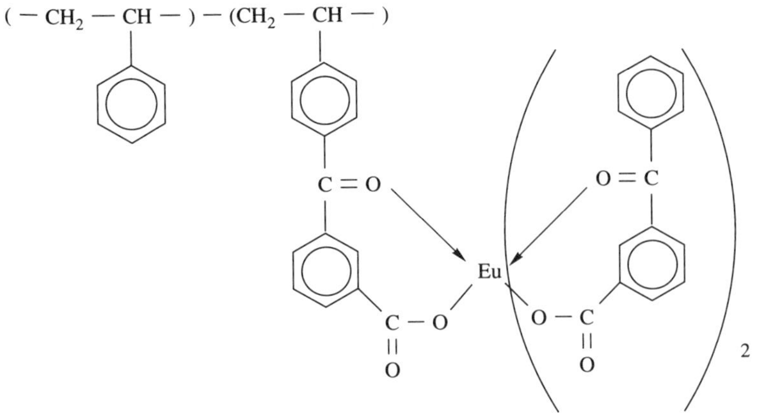

Still further, Eu(III) complex, 1-(*p*-vinylphenyl)-3-phenyl-1,3-propanedione, was copolymerized with MMA similarly.[381]

Radical copolymerization of a Co(II) complex containing *N,N*′-bis-(4-vinylbenzyl-oxy)salicyl aldehyde-1,2-diaminocyclohexane with styrene and divinylbenzene afforded a tercopolymer with one MCM unit per 20 comonomer units.[396]

A detailed consideration of the copolymerization of styrene, MMA, acrylonitrile (AN), and allylamine (AA) as well as chelate monomers such as Cu(II) complexes with pyrrolylmethyleneiminostyrene is given elsewhere:[383]

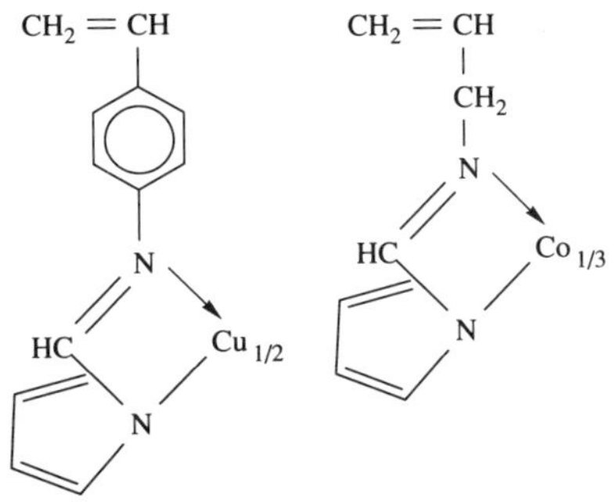

A mole fraction of monomer M_2 in the resulting copolymer is higher for the MCM than for its "metal-free" analogue. The monomer reactivity increases as it accepts a metal, a fairly rare situation in MCM polymerization conversions. A Co(III) chelate was copolymerized with the participation of no more than one or two multiple bonds out of three available. For a Cu(II) chelate the process involves typically both vinyl groups, but an initiation mechanism for the polymerization of this MCM is more intricate:

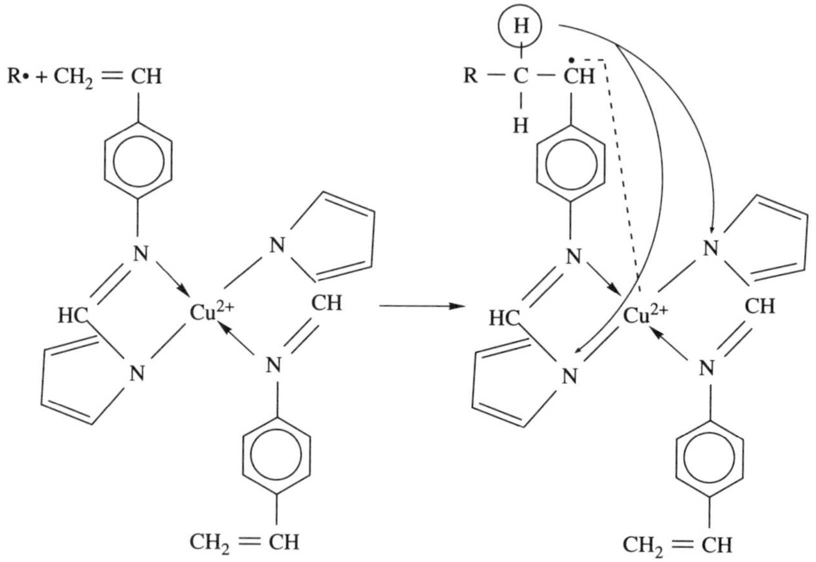

The primary radical attacks the vinyl group in the monomer first and then the free electron is transferred from this group to the central metal ion. This process is facilitated by the presence of a conjugation chain (a long coplanar π-electronic system). Chain propagation does not start until after a complete reduction of Cu(II) to Cu(I). A concurrent stage consists in β-elimination of a proton from the vinyl group attacked by the primary radical and its addition to the pyrrolyl nitrogen atom.

There are reports on the quantitative studies in the copolymerization of a Cu(II)-containing chelate monomer based on ethyl-α-(acetoacetoxymethyl)acrylate with styrene and benzyl methacrylate (40°C, THF, 2,2′-azo-2,4-dimethylvaleronitrile initiation).[369] The resulting copolymers contained as high as 52 and 72 mol% of MCMs (comonomers: styrene and benzyl methacrylate, respectively).

The copolymerization behavior of macrocyclic MCM chelates has been studied less thoroughly. Thus, monomers with high *Q* values (such as styrene, MMA, and π-conjugated species) can be effectively copolymerized with Fe(III)-containing dimethyl hemin ester, Fe(III) being reduced to Fe(II).[397] In the case of styrene, the content of the hemin derivative in the product is essentially consistent with the monomer mixture composition. As the [MCM]:[styrene] ratio increases, the yield, molecular mass (MM), and viscosity of the polymer decrease; this is attributed to the effect produced by metalloporphyrin on the chain transfer process (the chain transfer ratio = 2.3). As found for this MCM, Q = 70 and e = –0.17. At the same time, copolymerization that involves unconjugated monomers with a low *Q* value (*N*-vinylpyrrolidone [VPr], 1-vinylimidazole [VIA]) does not take place as the radical adds to hemin to afford stable products.

However, hemin copolymerization with participation of unconjugated monomers can be conducted in the presence of a third conjugated monomer to give, for example, a ternary hemin-VIA-styrene product.[397-399] There is evidence on the copolymerization of hemin with VPr or 2-hydroethyl methacrylate under ^{60}Co γ-irradiation in a buffer solution

Na_2CO_3–$NaHCO_3$, pH 10.0.[400] The MCMs obtained by interacting tetra-*p*-aminophenyl porphyrinate of Mn(II) acetate with acrylic chloride are readily copolymerized with MMA or 4–VP to give soluble products.[389] Tetra (*p*-vinyl)benzyl esters of a binuclear Mg(II) complex with chlorophyll are copolymerized with α-methylstyrene by the cationic mechanism (BF_3 etherate initiation).[401] Copolymerization of Co(II)-, Ni(II)-, and Fe(II)-containing vinylporphyrin monomers incorporating one or two side vinyl groups with acrylamide (AAm) or *N,N*-methylene-*bis*-acrylamide leads to the formation of a three-dimensional structure with simultaneous immobilization of MCMs inside the polymeric gel.[387] Note also the copolymerization of metallophthalocyanine monomers with 9-vinylcarbazole.[390]

D. GRAFT POLYMERIZATION OF CHELATE-TYPE MCMs

This area of polymerization MCM conversions is still at any early stage of development. Data concerning the grafting of chelated metal acrylates (methacroyl acetophenone, Dipy, and Phen are used as the chelating agents) onto a polyethylene (PE) powder, with preliminary irradiation in air, establishes a process significantly similar to this with the grafting of unchelated analogues in identical conditions (absence of homopolymers in the reaction products, presence of some unpolymerized vinyl groups in the graft copolymer, and like kinetic curves).[402-404] (See Figure 25.) Studies into the spectral and magnetic properties of the MCMs and their graft copolymers indicate that during the grafting the valence state and space surrounding of the metal ions remain unaltered, but the process is characterized by changes in the monomer-ligand (M–L) distance.

II. CLUSTER-TYPE MCMs

So far consideration has been given to polynuclear MCMs in which the central metal ion is surrounded by ligands of a given type incorporating polymerizable groups. Increasing the number of like atoms in a molecule gives rise to clusters. Cluster-type MCMs are

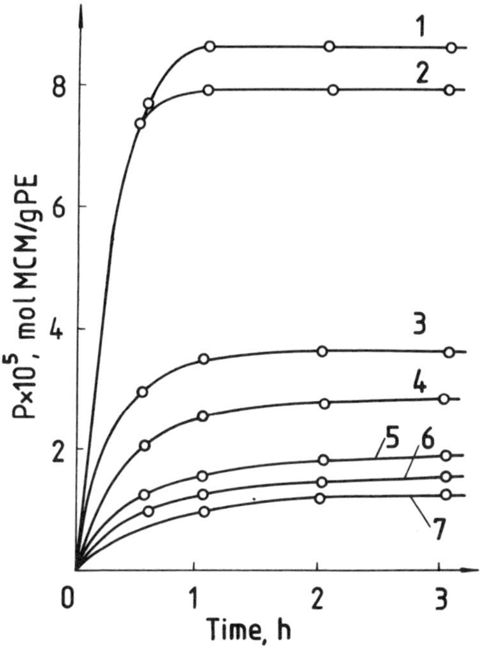

Figure 25 Kinetic curves for yield of graft polymers containing methacroyl acetophenone complexes: Co(II) (1), Ni(II) (2), Pd(II) (3), Mn(II) (4), Cu(II) (5), Fe(III) (6), and Cr(III) (7); ethyl acetate, [MCM] = 0.04 mol/l, 70°C, D = 200 kJ/kg.

those molecular compounds that include a frame made up of metal atoms spaced by short distances (<3.5 Å), which favor a direct M—M interaction, the frame being surrounded by polymerizable ligands.[405]

A. SYNTHESIS OF CLUSTER-TYPE MCMs

To do so, two major synthetic approaches are used: addition of polymerizable ligands to polynuclear complexes and completing the corresponding ligands with clusters.

The first approach can be exemplified by adding, in mild conditions, at the M=M bonds the following cluster:[406,407]

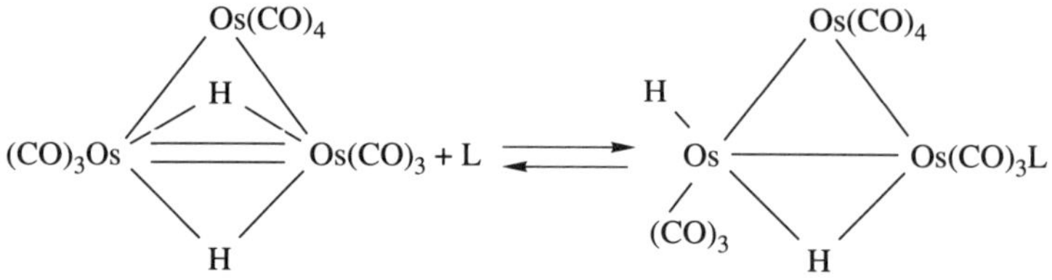

where L = $(CO)_3$, CH_3, or CN.

In this case the ligand can occupy both an axial and an equatorial position in the metal atom coordination sphere, and one of the hydride ligands passes from the bridge to the terminal type. However, as follows from the scheme below, the presence of the M=M bond is not a necessary condition for synthesizing this type of MCM:[408]

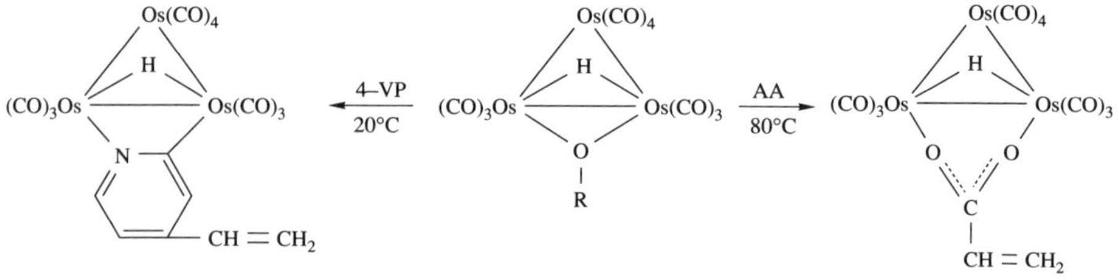

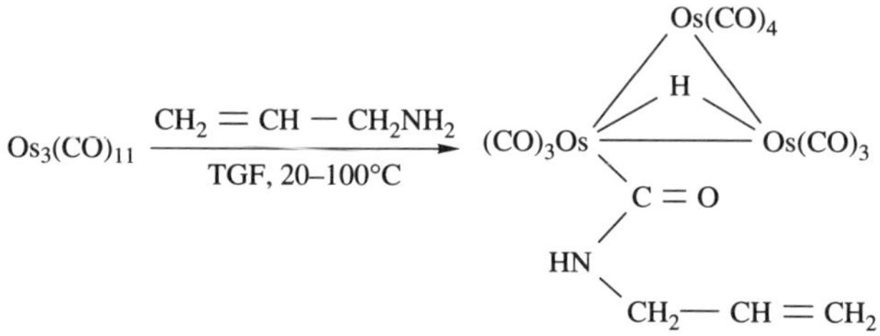

In the case of 4–VP, such MCMs are formed due to the orthometallation of the pyridine ring (with the appearance of an Os—C bond) and the *nV*-bond with the nitrogen atom. In other words, this is a kind of cluster-type chelate monomer. At the same time, the interaction of $Rh_4(CO)_{12}$ or $Rh_6(CO)_{16}$ with 4–VP is formed at the Rh ← N bond alone.

Also, the syntheses of cluster-type MCMs can be accompanied by an increase in the number of like atoms in the clusters, e.g., in the case of arenetetracobaltnonacarbonyl species:[409]

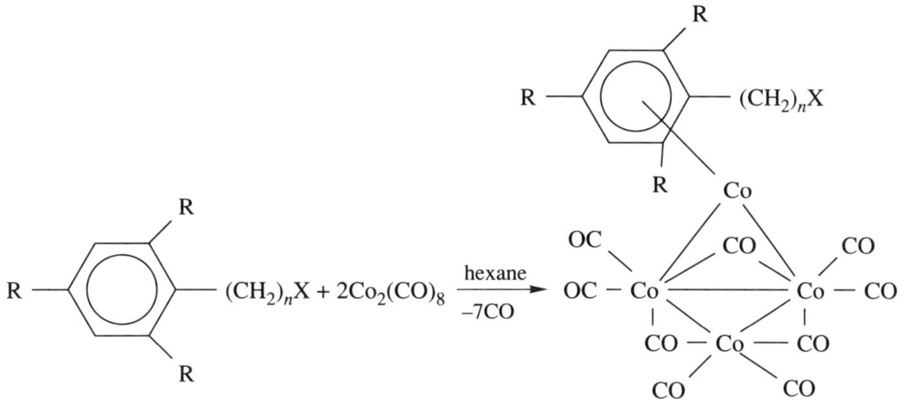

The second approach, i.e., completing a conventional-type monomer with clusters, is used less frequently. It can be illustrated by two characteristic cases. The interaction of $Co_2(CO)_8$ or $Fe_2(CO)_9$ with dithioesters (particularly, with methyl *p*-vinyldithiobenzoate), or with thioamides (*N*-cyclohexyl-4-vinylthiobenzamide) occurs as follows:[410,411]

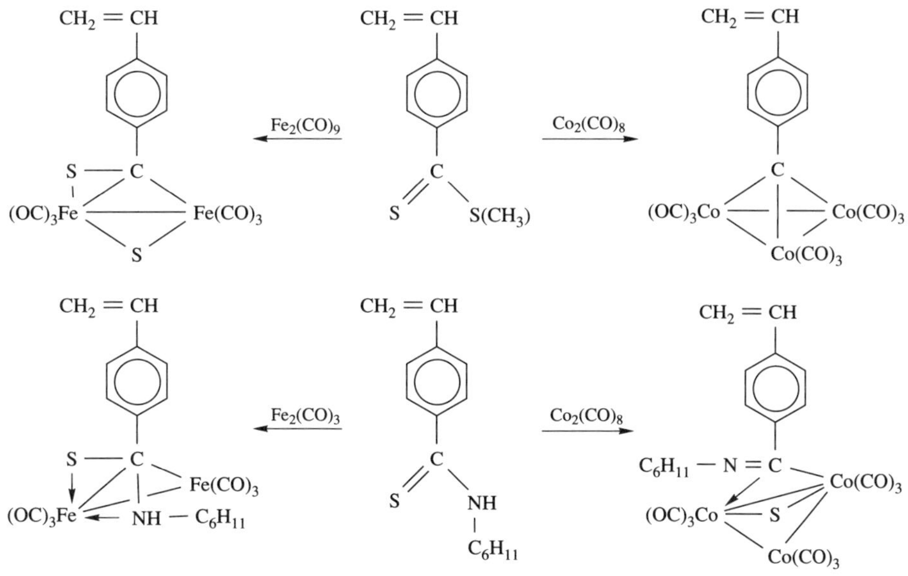

Such reactions are complicated by a number of side processes (desulfonation, as well as formation of metal sulfides and orthometallation products, etc.) with the result that the yield of cluster-type MCMs does not exceed 40%.

Table 15 **Composition and characteristics of cluster-containing copolymers**[410-412]

MCM (M_2)	M_1	Composition of monomeric mixture $M_1 : M_2$ (mol : mol)	MCM in copolymer m_2 (mol%)	M_w/M_n	Notes
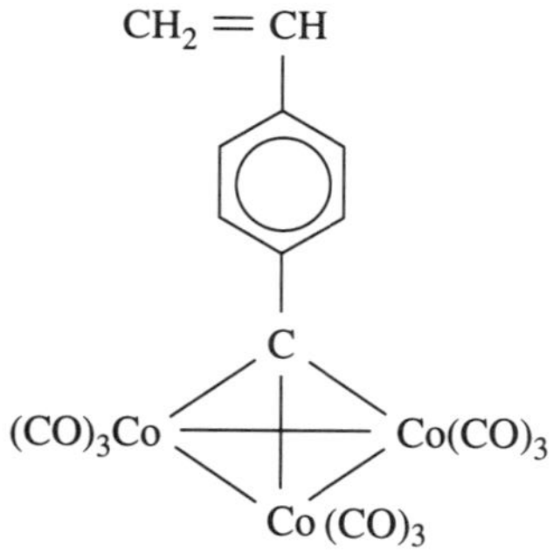	Styrene	99:1	3.7	14, 720/6,400	
	MMA	99:1	1.0	44, 700/21,300	
	N-tris-1,1,1-(hydroxy-methyl)methyl acrylamide	99:1	1.6		Water soluble
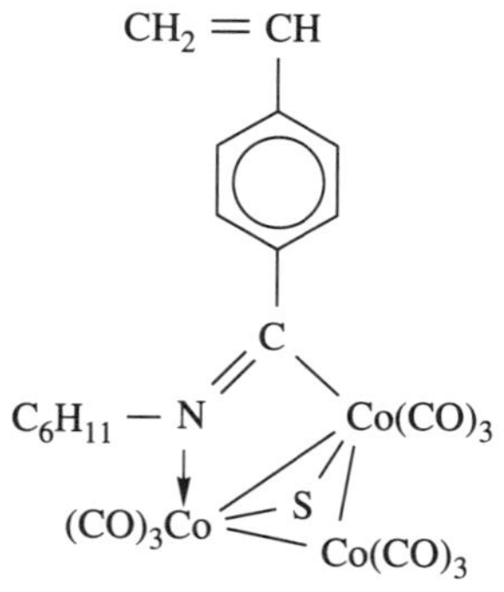	Styrene	99:1	1.3	16, 900/7,200	
	MMA	99:1	1.8	41, 300/24,600	
	N-tris-1,1,1-(hydroxy-methyl)methyl acrylamide	99:1	1.6		Water soluble

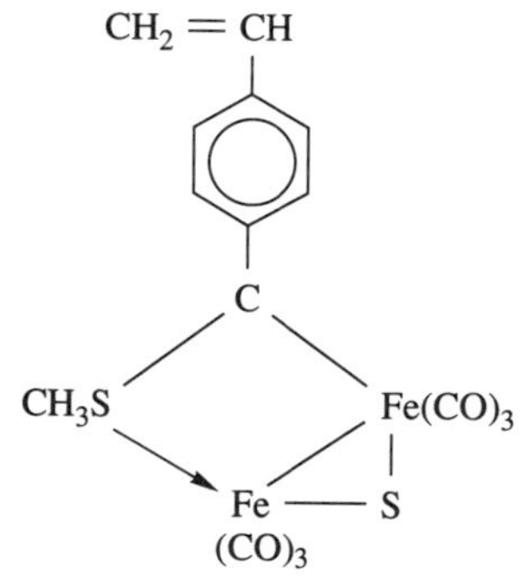	Styrene MMA *N*-tris-1,1,1-(hydroxy-methyl)methyl acrylamide	98.8:1.2 99:1 98.5:1.5	1.4 1.1 1.5	3,320/2,250 18,500/10,060 7,000/5,500	
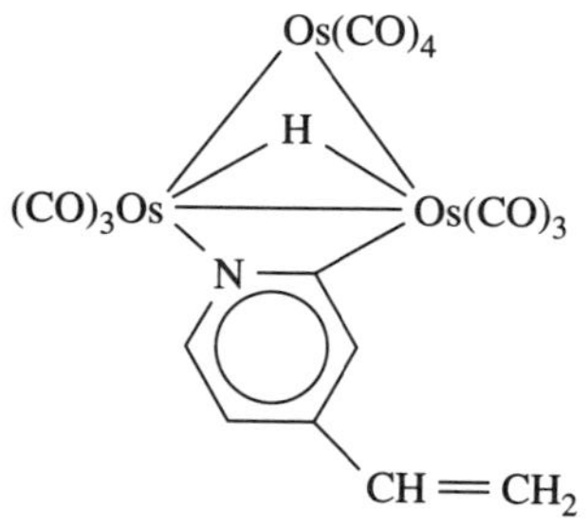	Styrene AN	99–99.5:0.5–1.0 99–99.5:0.5–1.0	0.2–1.0 0.1–0.2	—/5,000–36,500 —/15,000	In block In benzene

Table 15 (continued)

MCM (M_2)	M_1	Composition of monomeric mixture M_1 : M_2 (mol : mol)	MCM in copolymer m_2 (mol%)	M_w/M_n	Notes
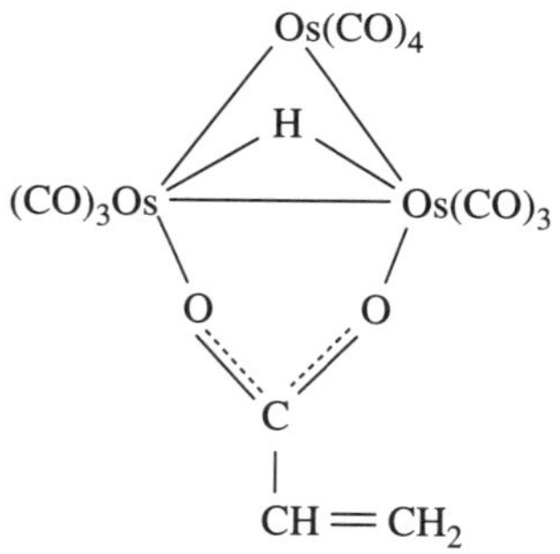	Styrene AN	99–99.5:0.5–1.0 99–99.9:0.1–1.0	0.1–1.0 0.1–1.0	—/12,000–25,500 —/8,000–26,000	In block In benzene
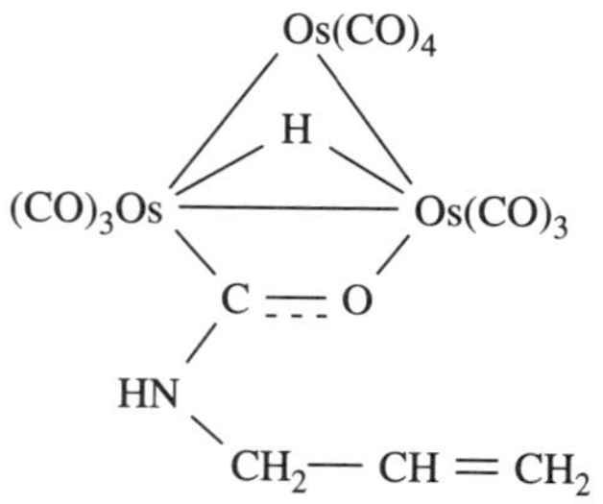	Styrene AN	99–99.5:0.5–1.0 99–99.5:0.5–1.0	0.1–1.0 0.1–1.0	—/21,000–26,000 —/15,000–17,000	In block In benzene

B. POLYMERIZATION CONVERSIONS OF CLUSTER-TYPE MCMs

Note that to the present time all attempts to homopolymerize cluster-type MCMs have not been successful. For the time being, the only way for making them participate in polymerization conversions is through copolymerization with conventional monomers. In this case, the units of the latter in the resulting products significantly predominate over those of the cluster-type MCMs. Thus, copolymerization of such Co(II)-containing MCMs with styrene or *N*-[*tris*-1,1(hydroxymethyl)methyl]acrylamide gave products incorporating as much as 1.5 mol% of clusters (Table 15).[410,411] Cluster integrity was proved by special studies. The content of Os clusters in copolymers with styrene and AN was even lower, 0.1 to 1.0 mol%.[412] Then, the presence of comparatively small amounts of a monomeric cluster led to a significant decrease in the copolymer MM. It should be pointed out that an alternative way for producing polymeric clusters, i.e., through interaction of clusters such as $Co_2(CO)_8$ with a polymer (a macroligand) involves some problems such as chain cross-linking, cluster adsorption, formation of colloidal metal species, etc. In other words, copolymerization of such MCMs is the best way for producing polymeric clusters characterized by high purity, product integrity, and structural homogeneity.

In summary, metallochelate and cluster-containing monomers represent a new generation of MCMs with problems needing further discussion. The same is true of heterometallic-type monomers, or MCMs including two or more dissimilar metals with a poor understanding of these species. For the time being, one can state that some progress has been made in metallochelate monomer syntheses and there are more modest achievements regarding MCM clusters. However, the polymerization properties of such MCMs have been mainly investigated on a qualitative level.

Chapter 5

Molecular and Structural Organization of Metal-Containing Polymers

In the preceding section when discussing the polymerization conversions of metal-containing monomers (MCMs), we were only interested in revealing the effect of a chemically bonded metal on the multiple bond reactivity. However, MCM copolymerization is undoubtedly interesting from another viewpoint, i.e., arrangement of the metal-containing polymers that appear: stereoregulation, behavior of macromolecules in solution and in the solid phase, and structural properties of the polyhedra. These problems are essential to finding relationships between the composition, structure, and properties of the products obtained by MCM polymerization.

I. STEREOREGULATION DURING MCM (CO)POLYMERIZATION

Polymerization of MCMs creates certain prerequisites for the appearance of regularly built polymers. They are based, first of all, on the possibility of the quadrupoles or ion pairs selecting one or another direction for approaching a monomer. Also, an important role is played by electrostatic and steric factors.

Usually, the microtacticity of the products resulting from the polymerization of metal methacrylates is established using a nuclear magnetic resonance (NMR) method after the methacrylate has been sequentially converted to polymethyl methacrylate (PMMA) (hydrolysis to PMAA and then methylation with azomethane).[126, 413] In the case of metal acrylate (co)polymers, they are first converted to (PAA) for which stereoregularity is analyzed by infrared (IR) spectroscopy. For syndiotactic PAA, absorption bands at 915, 963, and 1060/cm are typical,[414] and in an isotactic polymer spectrum the 1260/cm band splits into a doublet (1215 and 1275/cm).[415] In some specific cases the microtacticity can be assessed by fractionation, as well. Thus, the atactic PAA is soluble in water, whereas its syndiotactic variety is soluble in a dioxane-water mixture (80/20 vol%).[416]

Stereoregulation in the course of MCM polymerization is strongly dependent on the way the process is conducted. Thus, radical polymerization (benzene, benzoyl peroxide [BP], 70°C) was instrumental in producing sodium methacrylate in which the *syndio*- and *hetero*species were related in the range from 63.3:36.2 to 72.1:26.2.[186] At the same time, a reaction under anionic initiation conditions (C_4H_9MgBr, at temperatures between –5 and +2°C) gave rise mostly to an isotactic polymer (the content of the isotactic fraction was as high as 78.2 to 97.2%).

The possibility of stereoregulation during solid-phase polymerization was exemplified by Zn(II) acrylate (60°C, γ-radiation, 89.5°C).[157] The resulting polymer was less isotactic as compared with the polymerization product, $Ba[OCOC(CH_3)]_2 \cdot H_2O$, in similar conditions: at 100 and 50°C the contents of the *iso*-, *hetero*-, and *syndio*sequences were 49, 43, and 8%; 26, 48, and 26%, respectively. At the same time, the solid-phase postradiation polymerization of Ba(II) methacrylate dihydrate led to the formation of a less syndiotactic polymer as compared with that obtained by methacrylic acid (MAA) polymerization in solution.[69] The tacticity of that polymer depended appreciably on polymerization temperature and molecular mass (MM).

However, in some cases the resulting metal-containing polymers do not exhibit any preferred stereoregularity. Thus, the stereoregularity of the product resulting from tributyltin methacrylate (TBTM) polymerization (in block, 60°C, azobisisobutyronitrile [AIBN] initiation) resembled very much that of atactic polymethacrylic acid (PMAA).[417] This might be attributed to the block polymerization conditions, when the ionic pair effect demonstrates itself not as clearly as in the case when polymerization takes place in polar solvents. This can be corroborated by the results obtained in the polymerization of Co(II) and N(II) acrylates both in ethanol at 60°C (AIBN initiation) and in the case of a low-temperature polymerization on a glassy matrix defreezing when the polymers gave out as much as 65 wt.% of syndiotactic PAA.[146,156]

Formation of ion pairs composed of a growing radical and a counterion encourages syndiotactic addition.[126] Thus, MAA polymerization at high pH involves largely syndiotactically arranged units due to the electrostatic repulsion of the growing radical and the monomer.[418] At the same time in conditions when MCM dissociation is suppressed (e.g., in alcohol solutions), a higher role will be played by stereochemical factors, as well as the possibility of cation repulsion in the units.

Moreover, one has to remember that in the polymerization of bifunctional monomers such as Co(II), Ni(II), Mn(II), Ba(II), and other diacrylates the addition stereoregularity depends not only on such classical factors as the active site structure, temperature, and solvent type but also on the steric hindrances on the part of the appearing space-network structure of the metal-containing polymer.

The polymerization of such MCMs can be conveniently represented by two stages:

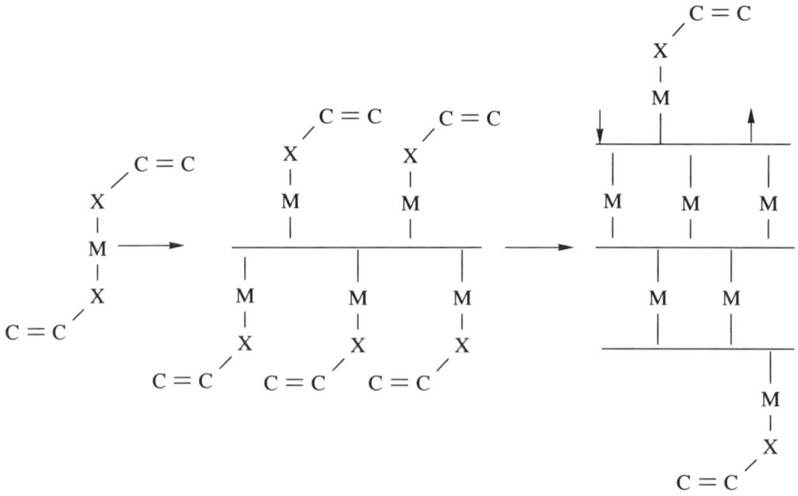

(1) (2)

Stage 1 is characterized by formation of a comblike linear polymer for which stereoregularity is defined by the classical factors. This stage probably gives rise to a stereoregulated polymer fraction. Stage 2 is responsible for the formation of the space-network structure proper. As this takes place, the polymer chain growth reaction mostly involves active sites and multiple bonds attached to the network skeleton. At this stage the chains grow in the presence of strong steric hindrances and an increasing level of internal (shrinkage) stresses to give an atactic chain structure. Also, these stages differ kinetically: the fast stage 1 is distinguished by a high conversion and the slow stage 2, by an ultimate conversion. Transition to stage 2 seems to be due to the deceleration of the chain growth reaction as a result of the restricted mobility of the C=C bonds attached to the skeleton of the appearing cross-linked structure and vitrification of the system.

A special stereocontrolling role is played by copolymerizations where MCMs are involved, the coordination effect being very important. The interaction arising between the approaching monomer and the segment units near the growing site might be responsible for configurational changes in the copolymer chains. In fact, the insertion of $[Co(NH_3)_5OCOC(CH_3){=}CH_2](ClO_4)_2$ in the copolymer molecule (especially at pH 3.0) leads to increasing content of the isotactic configurations.[88] The fact that the fraction of such configurations does not correlate with MCM content in the chain implies that a definite sequence of MCM units in the copolymer is not a necessary condition for the formation of isotactic configurations. Slight isotacticity increases in the copolymers appearing at pH 7.0 suggest that the effect of the terminal unit can be compensated for by electrostatic interactions between the oppositely charged MCM and MAA molecules. In the case of a Co(III) dimethacrylate derivative, the stereoregulating factor is not as pronounced since its methacrylate groups are located in a *cis*-position any of which favors addition reactions. On the other hand, in the monosubstituted monomer—$[Co(NH_3)_5OCOC(CH)_3{=}CH_2](ClO_4)_2$—the monomeric groups while being attached to the growing site are capable of selecting a definite direction for addition (due to the steric hindrances of the bulky side MCM groups).

It is reported that even small (up to 2 mol%) additions of transition metal acrylates to the copolymerizing methyl methacrylate (MMA)-MAA system (a molar ratio of 9:1) affect not only the relative reactivity constants of the comonomers, but also change the microtacticity of the copolymers.[184] This is especially pronounced during the formation of syndiotactic sequences composed of MMA units. The X-ray pictures of the modified copolymers exhibit distinct reflections characterizing the enhanced content of structurally ordered regions with denser macromolecule packing within the amorphous state.

Starting with the orienting influence of the coordination bonds and polar effects, one also can expect a significant impact of the terminal MCM unit on the structure of the attached unionized monomer. In fact in the copolymerization of butadiene and an organic alkoxyaluminum monomer, as the MCM fraction in the copolymer increases from 0.3 to 5.3 mol%, the fraction of the 1,4-*cis* units in the butadiene moiety microstructure of the copolymer decreases from 94.4 to 88.9%, this is due to the enhanced content of the 1,4-*trans* fraction (from 5.1 to 10.5%) at the same content of the 1,2-units (0.5 to 0.6).[26] It is believed that each MCM unit evenly distributed along the chain is preceded by a butadiene unit with a *trans*-structure resulting from the *cis*-isomerization of the active site in the insertion of butadiene.

Stereoregulation in the polymerization of the *nV*-type MCMs can be almost exclusively restricted to Zn(II) and Sn(IV) chloride complexes polymerized with MMA and MAN.[277,333] The stereoregularity of the metal-containing polymers obtained in some cases differs from that of the polymer generated from an uncoordinated ligand and changes appreciably with respect to the composition of the complex and metal type. Thus, the syndiotactic fraction of polymers obtained from $Zn(MMA)Cl_2$, $Zn(MMA)_2Cl_2$, MMA, $Sn(MMA)Cl_4$, or $Sn(MMA)_2Cl_4$ was 48, 60, 75, 77, or 63%, respectively. In this case, the liquid- and solid-phase polymerizations gave macrocomplexes of the same stereocomposition. Stereoregulation is markedly affected by polymerization temperature. As the temperature decreases, the content of the syndiotactic triads increases, whereas that of the heterotactic triads falls off at virtually the same contribution of the isotactic ones.

It should be noted that a difference in the activation enthalpy ($\delta\Delta H$) of *syndio*- and *iso*-additions of an uncoordinated MMA is 3.3 kJ/mol in the absence of substantial differences in the activation entropy ($\delta\Delta S$).[333] In contrast to this, in the polymerization of a complex-bonded MMA, the $\delta\Delta H$ value is close to zero and the probability of isotactic triad formation is only controlled by $\delta\Delta S$. This is attributed to the fact that MMA is fixed within a definite structure of the complex.

Formation of stereostructures also can be favored by a preliminary generation of charge-transfer complexes. This is normally accompanied by fixation of each pair of stereoconfigurations, for example, copolymers of the MMA-metal compound (MX_n)-styrene complexes.[294] It is evident that the number of syndiotactic diads is markedly smaller when $AlCl_3$ is used as MX_n to give strong complexes with MMA, which complicates an additional interaction with styrene.

Structuralization is also possible when macromers, particularly with MMA, were obtained.[11] Structural studies of the macromer indicated that it was predominantly isotactic, the syndiotacticity being higher when the polymer was synthesized from *p*-vinylbenzyl magnesium chloride than from an *m*-isomer and increased as the temperature decreased (at –110°C the triad syndiotacticity was as high as 96.6% at a low dispersion, $M_w/M_n = 1.19$).

It can be concluded that stereocontrol is also practicable in graft polymerization reactions. Thus, as suggested by IR spectroscopy of polyethylene (PE)-grafted PAA obtained after removal of the metal from the product resulting from transition metal acrylate graft polymerization, the *syndio*tactic structure of graft polymers is advantageous.[233]

II. MATRIX POLYMERIZATION OF MCMs

In the last few years a number of matrix MCM polymerizations have been reported. They state the fact that the "mother" polymer chain (a matrix) exercises a structural-chemical control over the elementary growth reactions with respect to the "daughter" chain arising from the matrix-adsorbed monomer macromolecules. As this takes place, MCM adsorption and association of the "daughter" chain fragments are effected through the chemical bonds with participation of the matrix units.

As to MCM polymerization, a special importance is attached to two matrix effects, namely, dynamic (the rate of matrix polymerization is usually higher than that of the ordinary one) and structural (orientation and nonvalent interactions of the matrix unit-sorbed MCM are responsible for stereoregulation in the "daughter" chain growth acts). In this context mention should be made of interesting studies of sodium methacrylate polymerization in aqueous solutions in the presence of chitosane acetate (a molar ratio of chitosane units to MCM is 1:1, $K_2S_2O_8$ initiation).[133] The rate of this reaction was described by the equation $w = K[M]^{0.8}[I]^0$ up to $[I] < 3.5 \cdot 10^{-3}$ mol/l and $w = K[M]^{0.8}[I]^{1.8}$ at $[I] > 3.5 \cdot 10^{-3}$ mol/l; whereas in the presence of a low molecular chitosane analogue, namely, glucosamine hydrochloride, it changed linearly with initiator concentration. The MM of sodium polymethacrylate changed linearly with conversion, the polymerization was distinguished by fast initiation and had a continuous form, and the polymer contained one SO_4 group per chain. All these facts corroborate the matrix nature of polymerization occurring through insertion of MCM molecules into the polymer chain, the molecules being associated with the chitosane units via ionic and hydrophobic bonds as follows:

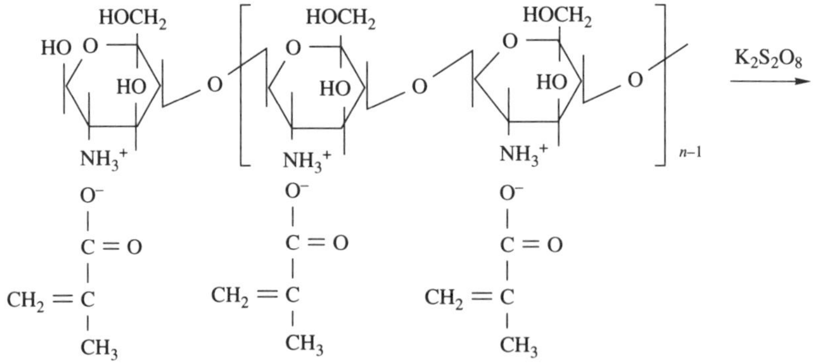

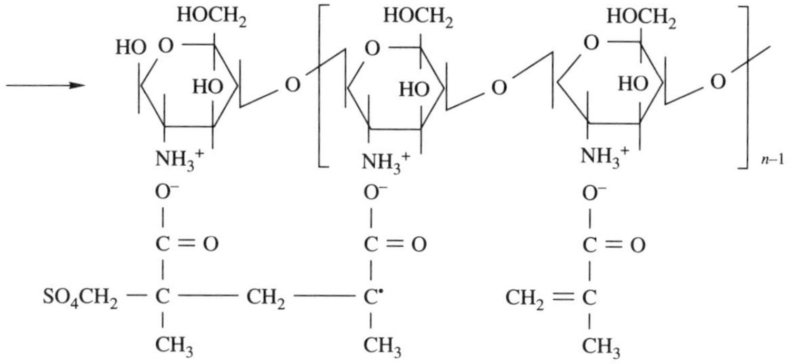

An ordered polyelectrolyte complex was also reported in the radical polymerization of sodium vinyl sulfonate in the presence of an ionene-hydroxide matrix.[419] The polymer complex retained the crystallinity of the parent monomers, but on dissolution of the macrocomplex in an NCl solution the resulting sodium vinyl sulfonate turned amorphous.

Special mention should be made of solid-phase photopolymerization of rigid species, namely, unsaturated organic tetrachlorocadmate cations, in the layers of halide-containing perovskite salts.[420] Such MCMs as cationic amidodiacetylene complexes of the type $(H_3NCH_2CO_2C{\equiv}CNH_3)_2CdCl_4$ held rigidly relative to the unsaturated side groups due to binding of the organic polymer with the inorganic layer form conjugated polymers. As this takes place, the inorganic salt layers act as matrices to give oriented polymers and the extra flexibility is conferred on the organic cation by binding it to the inorganic layer through one end only, rather than at both ends.

Polymerization of MCMs in a foreign polymer matrix used as the reaction environment offers great possibilities for modifying the properties of various materials. Thus, there are reports on the polymerization studies of acrylates and acrylamide (AAm) nitrate complexes of Zn(II) and Ni(II) in a high-density PE melt.[421] The MCM was added to the polymer from solution in methanol, and polymerization was conducted in a mold at 190°C and 15 MPa for 5 min. The addition of at least 0.5 wt% of Zn(II) acrylate was found to lead to the complete disappearance of its flow at 190°C. However, the addition of even 10 wt.% of MCM resulted only in a slight melting point (T_m) increase (from 131 to 135°C), the polymer crystallinity showing no changes at all. This suggests that in the course of polymerization the acrylate of the amorphous PE phase becomes cross-linked. Polymer cross-linking is characterized by a sharp drop of the relative elongation, by 790, 20, 10, and 2% on addition of 0, 2, 5 and 10 wt.% of Zn(II) acrylate, respectively; and the breaking strength of the modified polymer exhibits a maximum at 2 to 5 wt.% MCM (Figure 26).

Interesting findings were obtained in studies of laser-chemical polymerization of Ca(II), Co(II), Ni(II), and Cu(II) acrylates in a gelatin matrix.[422-424] Using metal acrylate additives, a recorder of holographic information with lasers can be made. The resulting holograms exhibit higher diffraction activities and resolutions as compared with conventional pictures. The process of gelatin cross-linking with Co(II) acrylate can be presented as follows:

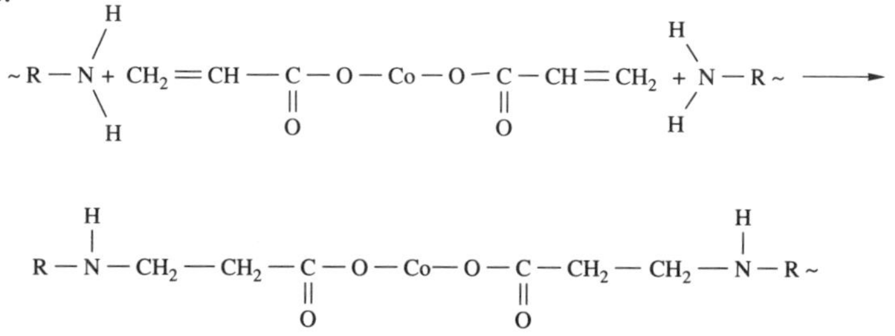

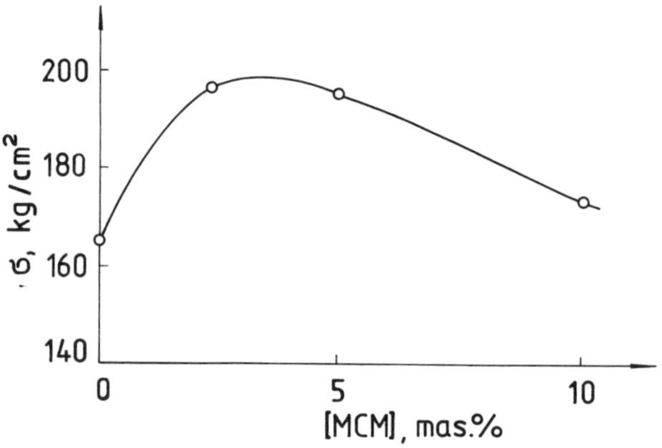

Figure 26 Tensile strength of Zn(II) acrylate polymerization product in high-density PE melt as a function of MCM content.

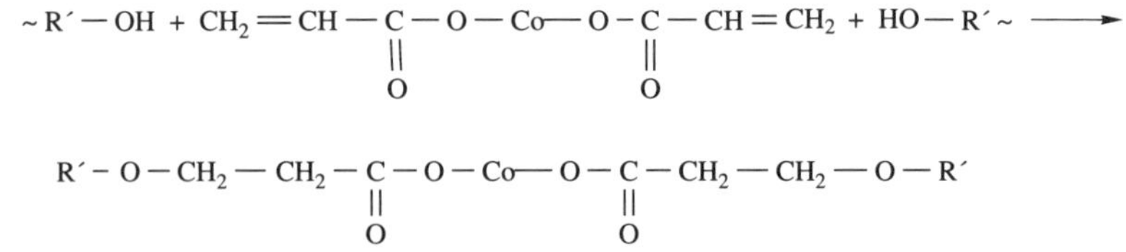

A similar process with participation of transition metal acrylates is more complicated and involves a preliminary formation of MCM and gelatin macrocomplexes, which is corroborated by IR and electron microscopy data. The process of photochemical structuralization of such complexes can be presented as follows:

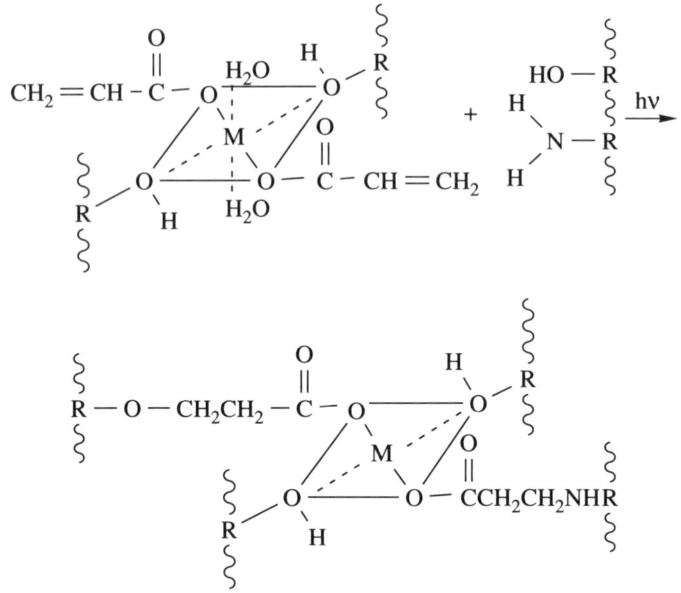

where R-NH is the macrocomplex fragment with argenine or lysine residues. The sensitivity of the system to the absorption of radiation by macrocomplexes depends on the nature of the transition metal used; the number of cross-links (or of the opened double bonds) changes similarly: Cu(II) > Ni(II) > Co(II).

III. PROPERTIES OF METAL-CONTAINING POLYMER SOLUTIONS

As in the case of conventional polymers, the properties of solutions of those species that result from MCM polymerization depend on the MM and geometry of the macromolecules, although the specific behavior of many of them is governed by the polyelectrolyte nature of the dissolved polymer.

The dissolution of metal-containing polymers is more often than not accompanied by dissociation, which is especially typical of polar solvents. In most cases this affects the geometry, the ligand surrounding, and sometimes the valent state of the metal. In aqueous solutions (especially at pH > 7.0) alkaline and alkaline-earth metal polyacrylates are almost in all cases dissociated, the dissociation being suppressed by adding nonpolymerizable salts (e.g., KCl). This can involve complex interactions. Thus, the heat of formation in sodium polyacrylate or a sodium acrylate copolymer with AAm-H_2O is negative, which is attributed to the polymeric salt dissociation; and the heat of its dissolution is positive due to hydrolysis.[425] In this case the concentration dependence of sodium polyacrylate diffusion coefficient (K_D) indicates that the effect of the macrocomplex charge on this quantity is more pronounced when samples with lower MM are employed.[426]

In ion-containing polymer solutions, the concentration dependence of the reduced viscosity in the presence of counterions decreases in the series $NH_4^+ > Na^+ > Mg^{2+} > Ca^{2+}$, which depends on the hydrodynamic interactions in aqueous and nonaqueous environments.[427] In fact, the preferred PMAA and sodium acrylate solvation in water-dioxane mixtures correlates with the solvent-induced conformational PMAA transition (at a 25 to 30 vol% dioxane content in the mixture).[428]

Of special interest is the temporal change of some properties of metal-containing polymer solutions. Thus, the viscosity of an aqueous sodium polyacrylate solution ($M_w = (4.5$ to $5.5)\cdot 10^6$, $M_w/M_n = 2.5$) drops in time within several weeks. However, in solutions with a high NaCl concentration this was not the case.[429] Such an effect is linked with a conformational change of the isolated macromolecules on the redistribution of the intramolecular hydrogen bonds and is interpreted as transition from a more rigid to a more flexible molecular structure. Moreover, the dissolution of the same metal-containing polymer with $M_w = 209{,}000$ was found to involve destructive processes; the result, in most cases, is the fact that the viscosity of the solution also decreases in time.[430] For dilute solutions (0.01%) one can observe two destruction stages: an initial fast (oxidative and mechanochemical) and a slower one, probably due to photolysis under ordinary lighting conditions. Under ultraviolet(UV) radiation sodium polyacrylate in dilute solutions undergoes a nearly complete decomposition to a monomer for no more than 1 h. The behavior of polymer solutions incorporating transition metals seems to be still more complicated, although no special studies have been made on this subject.

IV. STRUCTURAL ORGANIZATION OF METAL-CONTAINING POLYMERS IN THE SOLID PHASE

Many properties of metal-containing polymers are related to the formation of ion aggregates with a specific degree of association. Thus, the results of X-ray studies in aqueous sodium polyacrylate solutions suggest the existence of a certain order in the distribution of macroions even in solutions. One can observe a peak at small angles, corresponding to an interplane distance of 35 to 65 nm.[431] On the other hand, in the solid phase the metal-containing polymers show the presence of ion aggregates: pairs, multiplets (small compact ion groups), and clusters (aggregates of individual multiplets).[432,433] The contribution of these aggregates is assessed by a variety of physicochemical methods.[434] Thus in the diagrams showing the dependence of dielectric loss tangent on reverse temperature, the upper peak is assigned to the substance vitrification region formed by clusters and the lower one to the multiplet-containing material.[432] In the Raman spectra of sodium polyacrylate the absorption bands at 254 and 166/cm are due to ion vibrations in the multiplets and clusters, respectively.[435] The intensity of the latter correlates well with ion concentration in the clusters, as found by a dielectric method. Sodium salts of methacrylic, styrene-sulfonic, and styrene-*p*-carboxylic acid polymers were used to exemplify an important role played by the organic MCM moiety in that the position of the ion in the monomer affects the size or efficiency of the ion clusters, whereas the nature of ions controls the interaction strength therein.[436] In particular, depending on the ion type, the cluster-associated absorption band is observed at the following wavelengths: 250 ± 5/cm (Na^+), 185 ± 5/cm (Ba^{2+}), and 115 ± 5/cm (Cs^+). Basically, the structure of these aggregates can be presented as follows:[437]

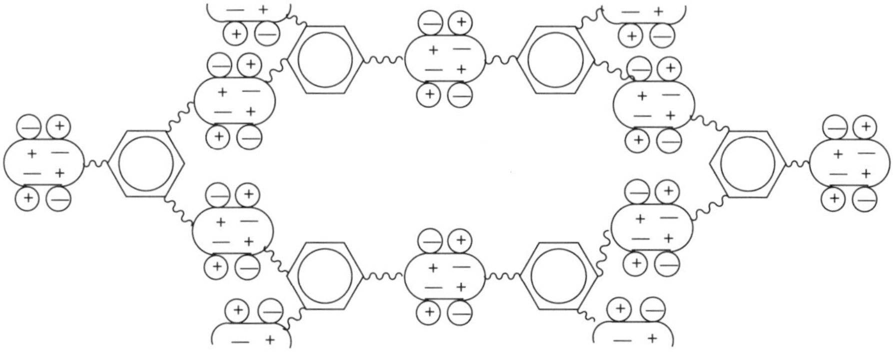

Usually clusters are assumed to be formed at high ion concentrations only (at least 8 wt%), although they have been found to exist in sodium polyacrylate copolymers in those samples alone where the total MCM concentration does not exceed 3.25%.[438] The structure of such ionomers depends on the dipole moments of the polymer. If there is a weak interaction between the polymer and the ions, they will be phase-separated to give ionic clusters. Thus, for a styrene-containing ionomer T_g changes by around 30°C but the dielectric constant remains unaltered to suggest that the dipole mobility amounts to the formation of ionic clusters.[439]

The cluster structure of ionic-type metal-containing polymers affects a number of physicomechanical characteristics such as modulus of relaxation, mechanical and dielectric loss tangents, etc. Thus the temperature dependence of the shear modulus for sodium polymethacrylate in the vicinity of T_g is a function of the T/T_g ratio; whereas on establishment of a certain critical temperature, the polymer flow pattern changes dramatically.[440] It is believed that this is associated with the decomposition of the multiplet structures made up of ion associates.

V. THE STATE OF TRANSITION METAL IONS IN METAL-CONTAINING COPOLYMERS

In the course of MCM polymerization conversions a change in the valent metal state is a rare case. A more frequent situation is when the coordination number and the shape of the complex polyhedron are affected. Thus, magnetic measurements corroborate the divalent state of the Co(II), Ni(II), and Cu(II) ions in the corresponding polyacrylates (Table 16).[146] The electron spectroscopy for chemical analysis (ESCA) spectrum of the Cu(II) polyacrylate shows the presence of a typical satellite (Figure 27) with an appearance that is due to the availability of a lone electron in the metal ion. As opposed to the MCM, the polymer is distinguished by a lower intensity of the satellite and a narrower half width of the base line which shifts to the low-binding energy side (935.0 →

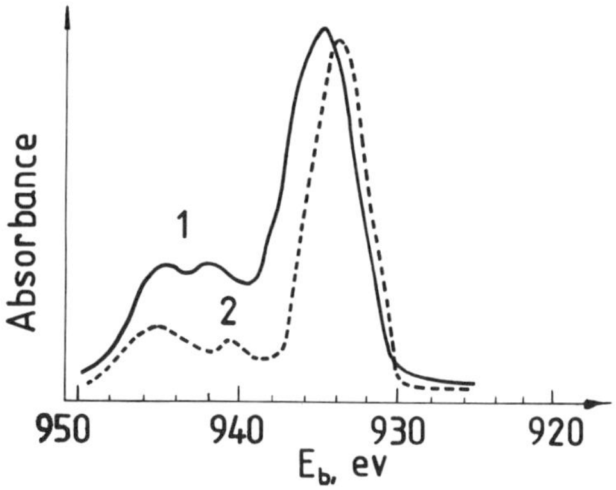

Figure 27 ESCA spectra ($2p_{3/2}$) of Cu(II) acrylate (1) and polyacrylate (2).

Table 16 **Magnetic properties of transition metal polyacrylates**[146]

	μ_{ef}, μ_B	
Metal ion	**22°C**	**–196°C**
Ni(II)	5.05	4.52
Cu(II)	3.60	3.47
Co(II)	1.19	1.05

934.0 eV). The ESCA data indicate the presence of small amounts of Cu(I) ions in both a homo- and a copolymer of the Cu(II) acrylate with Ti(IV) dicyclopentadienyl dimethacrylate.[168] According to the electron spin resonance (ESR) evidence, in these MCMs at a low copper content the complexes have the form of an extended octahedron.

Although the polymerization of Co(II) and Ni(II) acrylates does not involve any significant changes in the coordination metal ion environment, the nepheloxetic parameter β for the polyacrylates is larger than that for the corresponding MCMs (0.897 and 0.860 against 0.886 and 0.839, respectively). This suggests a certain drop of the orbital contribution to the metal-ligand σ-bond in the macromolecules. As with alkali metal ions, their transition metal counterparts included in the macrochain can form various associates. However, more often they are scattered along the chain as, for instance, in the case of Co(II) and Ni(II) polyacrylates obtained by radical MCM polymerization in solution.[146] For Cu(II) polyacrylate one can observe exchange interactions probably due to a specific binuclear structure of the parent MCM.

At the same time, in metal-containing copolymers based on transition metal acrylates and obtained in the solid phase under high pressure and shear strains, one can observe a vigorous antiferromagnetic exchange between the paramagnetic sites. Its appearance is likely due to the interchain interactions of these sites as a result of conformational changes in the macrochains during the process. The fact that no exchange takes place in finished metal-containing copolymers produced in solutions and exposed to high pressure and shear strains suggests that the structure of the complexes associated into clusters through antiferromagnetic interactions is formed at the stage of polymerization in the solid phase.

In copolymers composed of different metals and incorporating diamagnetic ions, the paramagnetic sites can be isolated with the diamagnetic site units. Such was the case with the copolymers of Cu(II) and Ti(IV),[68] and Fe(III) and Zn(II)[220] acrylates. However, even the remote groups produce a strong inductive effect on the electron levels and polarizability of the functional groups in such polymers.

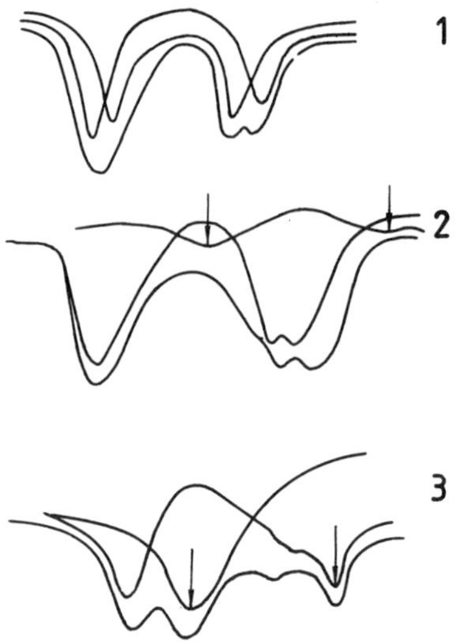

Figure 28 Mössbauer spectra of MMA copolymer with iron cluster at –269°C (1), –196°C (2), and 7°C (3).

Moreover, ESR, Mössbauer, and ESCA spectroscopic studies—as well as magnetic measurements of graft metal-containing polymers based on metal acrylates and AAm complexes—indicate that the valent and coordination states of the metal atoms remain unaltered.[233] The same is true of the valent state of Co(II) in the graft polyallyl alcohol-$CoCl_2$ complex.[358] At the same time, the magnitude of magnetic moment of Co(II) (5.3 μ_B) suggests that the tetrahedral surrounding of this metal in the monomeric complex changes to an octahedral one in the graft polymer.

Interesting features are exhibited by the structure of copolymers based on cluster-type MCMs. Mössbauer studies of polymeric iron-containing clusters (Figure 28) disclose a doublet characteristic of an iron oxide doublet (designated by the arrow).[441] The isomeric shift and quadrupole splitting magnitudes indicate that the electron state of the iron atoms in the MCM and in the copolymer are the same, whereas the chemically modified polymers using $Fe_2(CO)_9$ show the presence of various products corresponding to the appropriate methods of iron coordination. This corroborates once again that the polymerization conversion of the MCMs under study is the most convenient way for producing structurally homogeneous polymers incorporating cluster groups.

In conclusion, the structural organization of metal-containing copolymers is exercised at all levels: molecular—pertaining to the chemical structure of the polymer units, metal distribution in the chain, and stereochemical structure of the chains themselves; supramolecular—reflecting the intermolecular interactions, the degree of macromolecule ordering and packing; and topological—characterizing the degree to which the polymer structure elements are associated. An important role is played by the structural organization of the metals themselves; and the processes of their clustering await further studies. The structural organization of metal-containing polymers exerts a great effect on their principal properties.

Chapter 6

Special Properties and Applications of Metal-Containing Polymers

Materials resulting from polymerization metal-containing polymer (MCM) conversions possess a variety of different, sometimes unique, properties which predetermine their applications. These properties can stem directly from the presence of a metal (e.g., catalytic and biological activities) or from the structural MCM characteristics (enhancement of resistance, physicomechanical and operating characteristics, etc.).

I. CATALYSIS WITH THE USE OF METAL-CONTAINING MONOMERS AND POLYMERS

The subject MCMs and the associated polymers can act as homogeneous or pseudo-homogeneous catalysts of a variety of processes such as polymerization, hydration, oxidation, hydroformylation, etc. At the same time, copolymers afforded by MCMs represent one of the species of immobilized catalysts, or metallocomplexes, fixed on polymers that play the role of macroligands. The structure of such catalysts is distinguished by a controlled arrangement of a metal along the chain with the possibility of identifying the metallocomplex structure more closely. As a rule, such catalysts inherit the advantages of conventionally produced macrocomplexes, such as high activity and selectivity, stability at enhanced temperature, ease of removing the reaction product, and catalyst regeneration.

A. INITIATION OF POLYMERIZATION REACTIONS

Vinyl monomers are efficiently polymerized with styrene-sodium acrylate or MM-styrene sodium methacrylate copolymers in aqueous environments at 85°C.[442] The use of Zr(IV) methacrylate, $Zr_4(OCOC(CH_3){=}CH_2)_{10}O_2X_2{\cdot}H_2O$, was proposed (where X = OH^-, $CH_2{=}C(CH_3)OCO^-$, etc. and n = 2 or 4) as a catalyst for the radical block polymerization of vinyl monomers.[443] A covalent-type MCM, *m*-$CH_2{=}CHC_6H_4CH_2MgCl$, is a good catalyst suitable for producing polymethyl methacrylate (PMMA) with a high syndiotacticity and a narrow molecular mass distribution,[11] and potassium acrylate for initiating the polymerization of β-propiollactone in the presence of crown ethers.[444]

Another interesting example is vinylpyridine polymerization in methanol-pyridine media, initiated with a $Cu(4{-}VP)_2(OCOCH_3)_2$ complex.[445] The general equation for the process rate is as follows:

$$w = K[Cu(II)]^{1/2}\,[4{-}VP]^2$$

at E_a = 50 kJ/mol, against E_a = 113 kJ/mol in the case of azobisisobutyronitrile (AIBN) initiation. The polymerization is initiated by transferring an electron from the monomer to the copper ion in the complex:

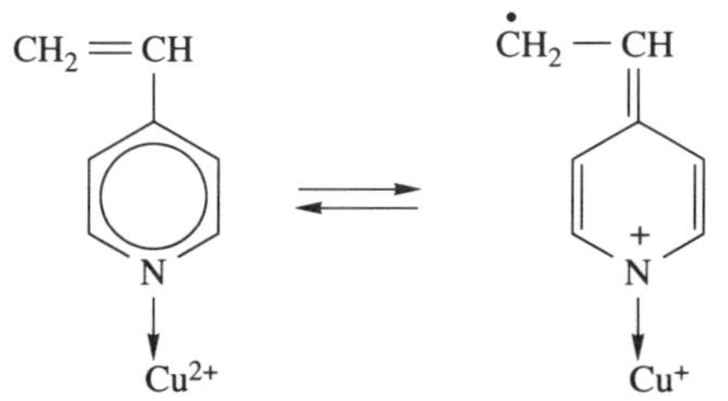

This process is distinguished by the fact that the electron is transferred from the multiple bond π-electron rather than from the free radical as is the case of Cu(II)-containing monomers. This leads to a limiting case of its activation, or opening. The initiator is reactivated by oxidation of the resulting Cu(I) to Cu(II) in air.

Note also the catalytic activity of $ZnCl_2$-vinylimidazole complexes when epoxy resin is cured at 120 to 170°C.[446] The resulting compositions exhibited high heat resistance and favorable physicomechanical characteristics.

The products of MCM conversions are applied to polymerization processes less often than MCMs themselves. In this situation the only case to be mentioned is the use of an acrylamide (AAm) copolymer with *N,N*-methylene bisacrylamide and a Ni(II) vinylporphyrin complex for creating efficient catalytic systems (in combination with H_2O_2) to initiate AAm polymerizations in aqueous media.[387]

B. OLEFIN POLYMERIZATION

In the analysis of anionic-coordination MCM copolymerization with olefins and dienes, it was mentioned that it is possible to use MCMs for creating stable and high-active catalytic systems for ethylene polymerization (Figure 29). Their persistent action is caused by the heterogeneous nature of the system in that the formation and destruction of the active sites is decelerated due to the fact that they appear only on the surfaces of complexes insoluble in the reaction environment—metal compound (MX_n) with vinylpyridine (VP), methyl methacrylate (MMA), acrylonitrile (AN), etc.[447] Another important factor is the additional possibility of controlling properties of the resulting polymers.[240,270,271,347] Thus, in propylene polymerization a $TiCl_3$-methylacrylate (MA) complex increases the catalytic activity of the system by a factor of ca. 2 and enhances its stereospecificity.[383] Such MCMs as Ti(IV) alkoxy derivatives, as well as the associated homo- and copolymers, are also active components of ethylene polymerization catalysts.[448] Homopolymers of Ni(II) complexes with dipyridyl (Dipy) or 1,10-phenanthroline (Phen) effectively catalyze the process of ethylene dimerization.[449]

The catalytic activity of metal-containing polymers shows a high increase when the reaction sites are brought out to the surface, which is effected by means of graft polymerization. Thus on the surface of powdered polyethylene (PE), polypropylene (PP), or polystyrene, graft copolymers of allyloxytripropoxytitanium, allyltitanocene chloride, and other compounds are far more active in ethylene copolymerization as compared with the products of their homo- and copolymerization.[35] Thus, a PE-graft Ni(II) polyacrylate or a polyacrylamide (PAAm) complex is a good catalyst for ethylene dimerization in a

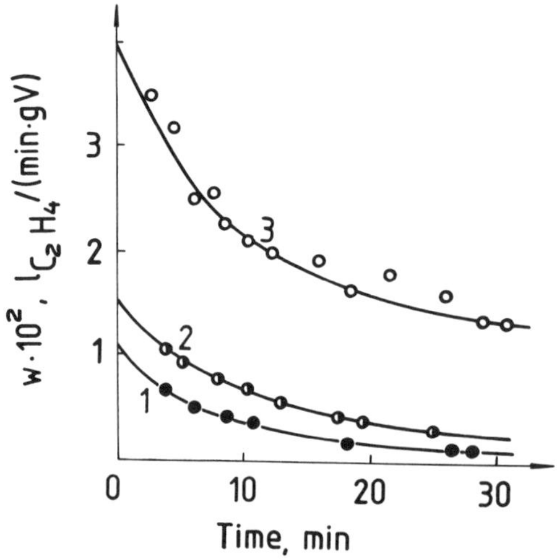

Figure 29 Kinetics of catalytic ethylene copolymerization: $V(4\text{-}VP)_2Cl_4\text{-}(iso\text{-}C_4H_9)_2AlCl$ (heptane) (1), $V(4\text{-}VP)_2Cl_4\text{-}SnCl_4$ ([Sn]:[V] = 1.5)-$(iso\text{-}C_4H_9)_2AlCl$; (heptane) (3), $V(4\text{-}VP)_2Cl_4\text{-}(iso\text{-}C_4H_9)_2AlCl$ (toluene) (2); $[C_2H_4]$ = 0.026 mol/l, $[V(4\text{-}VP)_2Cl_4]$ = 1.7 mmol/l, [Al]/[V] = 14, solvent = 100 ml.

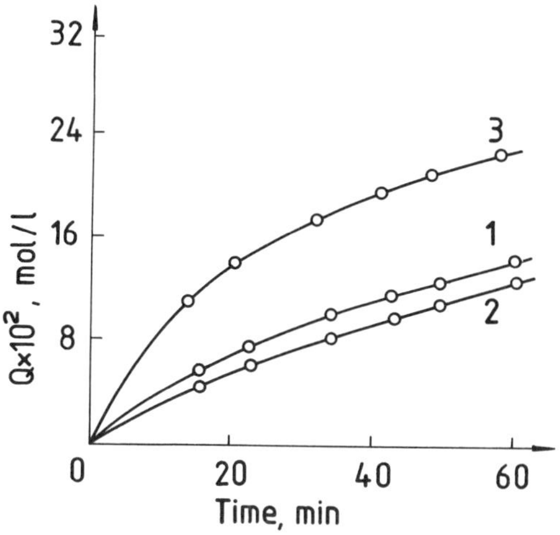

Figure 30 Buildup of butene isomers during ethylene dimerization in catalytic system: Ni(II) polyacrylate grafted onto PE-Al(C_2H_5)$_2$Cl: butene-1 (1), butene-2-*cis* (2), butene-2-*trans* (3); C_2H_4 pressure = 0.4 MPa, heptane, 70°C.

mixture of isomeric butenes (Figure 30).[450] The same is true of rubber-graft complexes composed of Ni(II) compounds and VP.[355]

The trend has been toward the use of bimetallic systems containing atoms with catalytically differing functions. Thus, such a catalyst as a copolymer based on a N(II) and Ti(IV) acrylate was instrumental in using a "relay-race"-type ethylene copolymerization in which the conversion product appearing on one of the sites (namely, butene on Ni(II)) serves as a substrate for the second site, Ti(IV), responsible for its polymerization with ethylene. Such catalysts are very promising due to the possibility of building systems with controlled distances between the various reaction sites.

C. HYDROGENATION OF UNSATURATED SUBSTRATES

Essentially all MCMs and their polymers containing noble and some other metals are capable (after being preactivated, e.g., with $NaBH_4$, hydrazine, and hydrogen) of activating the hydrogen as it adds to the multiple bonds. Thus, mention can be made of polyvinyltitanocene chloride catalyzing the hydrogenation of unsaturated hydrocarbons.[452] Its activity is comparable with that of polystyrene immobolized $(C_2H_5)_2TiCl_2$. Polystyrene-dipyridylpalladium gives good results in the conversion of 1,3-hexadiene to a mixture of hexenes.[453]

A high selectivity is displayed by the reductive amination of nitrobenzene and the products of its reduction with $PdCl_2$ complexes composed of 4-vinylpyridine (4–VP) and other PE-graft ligands with nitrobenzene compounds.[454] The product resulting from the graft polymerization of Pd(II) chelate with a diketone ligand proved efficient in hydrogenation chloronitrobenzene isomers.[455] This catalyst retains its activity over five cycles.

D. HYDROCARBON OXIDATION

Extensive works dealing with the oxidation of unsaturated hydrocarbons are under way at present, since this is one of the ways for producing a series of valuable oxygen-containing products. The key aspect of this problem is constituted by selectivity of the process. In the case of some metallopolymeric catalysts, hydrocarbons in the liquid phase are oxidized (as usual) by the chain mechanism with the participation of free radicals: the major stages are chain initiation, propagation and branching; chain and molecular hydroperoxide decomposition; and finally chain termination.[456,457] In the model reaction of cyclohexene oxidation in the presence of a Ni(II) acrylate-styrene copolymer as much as 90% of the resulting product is represented by cyclohexenylhydroperoxide (Figure 31). At the same time, dimethylvinylethynylcarbinoltributoxytitanium does not exhibit any

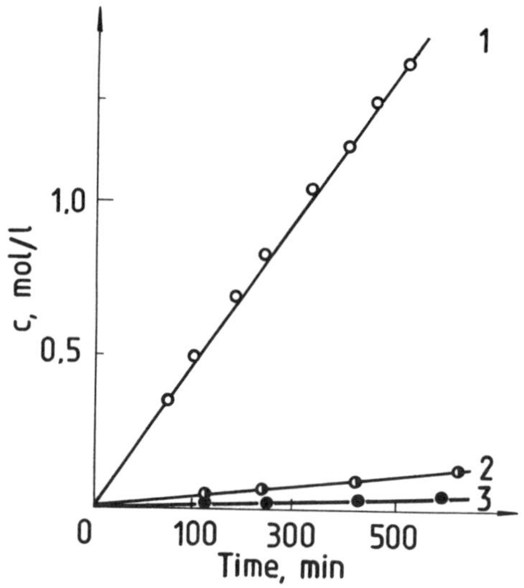

Figure 31 Kinetics of cyclohexene oxidation with molecular oxygen in the presence of catalyst Ni(II) acrylate copolymer with styrene. Oxidation products: cyclohexenyl hydroperoxide (1), cyclohexenol + cyclohexenone (2), cyclohexene oxide (3); Ni(II) content in copolymer 64 mol% [Ni] = 4 mmol/l, O_2 pressure = 0.1 MPa, 60°C.

selectivity in this process, displays a clear-cut epoxidizing ability, and concurrently facilitates hydroperoxide formation. When applying a PE-graft Co(II) polyacrylate to this reaction, the process is characterized by a high steady rate exceeding that of the oxidation which is catalyzed with the immobilized PE-graft (PAA)-Co(II) system by a factor of 2. This might be associated with a difference in the mechanism of reaction initiation due to the diverse Co(II) ligand surrounding. It should be noted that metallopolymeric catalysts can be used in more than five cycles.

Other metal-containing polymeric catalysts can be exemplified by soluble metalloporphyrins produced by MMA or 4–VP copolymerization with a macrocyclic metallochelate monomer resulting from the interaction of allylamine (AA) chloride with Mn(II) acetate tetra-*p*-amino-phenylporphyrinate, which acts as efficient catalysts of olefin oxidation with molecular oxygen.[458]

E. PHOTOCHEMICAL REACTIONS

Recent years have been marked by an ever growing application of metal-containing copolymers to chemical solar energy conversion systems, and primarily those designed for photocatalytic water decomposition to O_2 and H_2:[459]

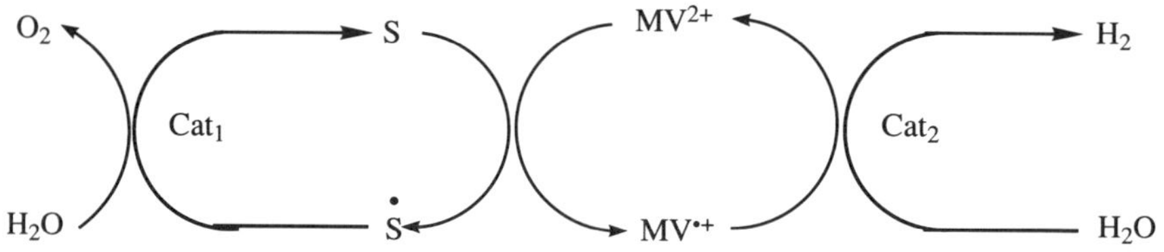

Here Cat_1 and Cat_2 are the catalysts for O_2 and H_2 separation (clusters including cobalt, iron, copper, manganese, and platinum metal colloids); S is the sensitizer (photocatalyst), and MV^{2+} (methylviologene) is the reducing (electron-transfer) agent.

Metal-containing copolymers, particularly those comprising MCMs which incorporate trispyridyl Ru(III) complexes, are used as sensitizers:

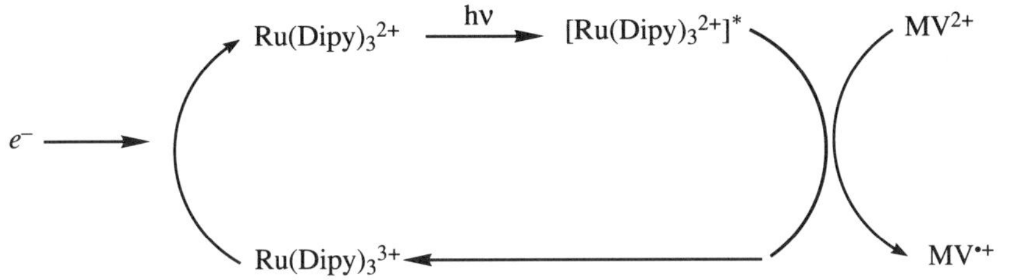

Then Ru(III) ("a hole") oxidizes Cat_1 to the high valent states of the ions, (namely, Co(IV), Fe(IV), Ni(IV), Cu(III), and Mn(IV)) which during a synchronous four-electron process liberate O_2 from two water molecules. Thus, there are reports on highly efficient water-soluble catalysts of water photodecomposition, such as copolymers of *N*-ethylated VP and *cis*-di(2,2´-dipyridyl) (4-methyl-4´-vinyl-2,2´-dipyridyl) ruthenuim(II).[460] In this case, the use of electrodes coated with films made of such a metallopolymer accelerates the photochemical reaction due to the fact that the reverse process (i.e., electron transfer from $MV^{\bullet+}$ to $Ru(Dipy)_3^{3+}$) will be impeded.[461]

Another alternative—instead of using sensitizing $Ru(Dipy)_3^{3+}$-containing polymers—involves the use of polymeric macrocyclic Mg(II), Mn(II), Fe(III), and Ru(II) metallocomplexes of the phthalocyanine, salene, and other types.[462] In many respects they are similar to natural sensitizers, namely, chlorophyll. The same metallopolymers are very suitable for a photosensitized conversion of norbornadiene to tetracycline coupled with a catalytic endothermic opposite conversion under the effect of stored solar energy in the presence of a catalyst.

Note also that protoporphyrin IX monomer displays activity in the photoreduction (xenon lamp irradiation) of red stable A with L-ascorbic acid. The same is true of benzoquinone reduction in aqueous and alcohol-aqueous environments.[463] On the other hand, the addition of Mg(II) to the vinylporphyrin cycle amplifies the catalytic properties of this system.

F. OTHER CATALYTIC REACTIONS

The applications of MCMs and associated polymers are not restricted to the above examples. Thus, polymers containing styryl-phosphine Co(II), Ni(II), and Pd(II) complexes act as catalysts in olefin hydroformylation reactions.[464] In addition, $PdCl_2$-vinylimidazole complexes have been found to exhibit a high activity in phenylacetylene hydroxylation:[465]

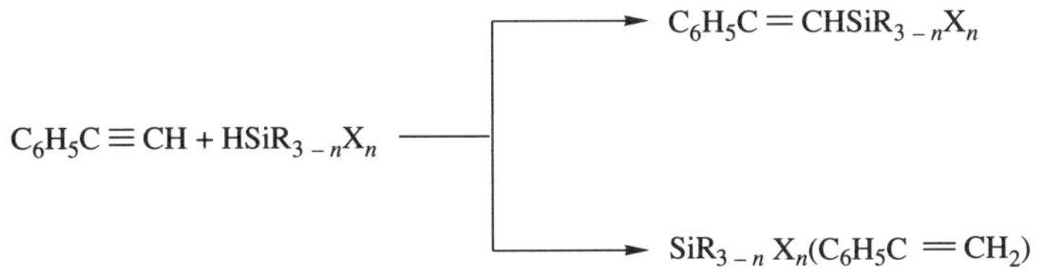

The product yield was as high as 98% and the ratio between the isomers, 0.02.

II. BIOLOGICAL ACTIVITY OF METAL-CONTAINING (CO)POLYMERS

Polymers based on heavy metals and primarily tin have been used to advantage for the production of antifouling coatings in which the pesticide groups are chemically linked with the main polymer chain.[7,466] Such coatings exert a significant prolongation effect, are ready to apply, and strongly adhere to the base. In addition, they can be used as structural materials.

The biological mechanism of such polymers is due to the fact that in the presence of water and enzymes liberated by microorganisms responsible for fouling, polymeric biocides emanate strongly toxic low molecular compounds that affect these microorganisms.[467] As this takes place, the toxin is leached from the polymer at extremely low rates, around 0.004 mg/cm^2/d, which is two orders of magnitude lower than in the case of low molecular organotin compounds. This provides for the protection of the products over 3 to 5 years of operation.

The favorable biological effect of metal-containing polymers is due to the changing nature of an organotin biocide, in this case an R_3Sn^+ cation:

$$-CH_2-\underset{\substack{| \\ COOSnR_3}}{CH}- \longrightarrow -CH_2-\underset{\substack{| \\ COO:^-}}{CH}- + R_3Sn^+$$

Organotin cations appear to block the sulfur-containing protein groups responsible for the formation of a ternary protein structure, which affects the chain conformation and leads to a complete or a partial protein denaturation. These processes depend markedly on the nature of the metal, type of the biocide bonded to the polymer base, and the structure of the MCM unit. Thus, trialkylmethacrylplumbanes are more active with respect to daphnia than to algae, in that they stop their growth and reproducibility even at concentrations of 0.005 mg Pb per liter. Biotoxicity also can be controlled by the degree of copolymer cross-linking: the higher this parameter, the lower the toxin-leaching rate.

The biological activity of sulfur-containing polymers varies widely: they display fungicide, insecticide, and antivirus activities. At present, a great many compositions are known; however, the highest activity seems to be exhibited by tin-containing maleic anhydride copolymers, polytrialkyl(aryl)tin acrylates, maleates, and itaconates.

As with catalysis, the biological protection of materials requires a surface distribution of the corresponding compounds. In this context, a great interest is generated in graft copolymers based on MCMs that incorporate biocide groups. To impart a suitable bioresistance to PE, PP, polyamide, poly(ethylene terephthalate) (PETP), polyvinyl chloride (PVC), and wood, it would be sufficient to have no more than 0.5 to 1% of graft polytriethyltin methacrylate with a significantly extended time of biological activity.[230] Also, tin-containing polymers can be used in various combinations to protect fibers, paper, textile, and other materials.[468]

Note also the fungicidal activity of a mercury-containing polymer product such as an MMA copolymer with 0.1 to 0.4% of Hg(II) methacrylate.[469]

The biocidal activity of some copolymers based on sodium acrylate has been reported.[470] This activity grows on the addition of sulfonate groups of another MCM, namely, sodium vinyl sulfonate. One more sodium-containing monomer (its sulfohexyl methacrylate) displays an anticoagulating activity with respect to blood,[186] whereas an iron acrylate-AA copolymer (feracryl) is used to arrest bleeding.[471]

The biological activity of metal-containing polymers can be further exemplified by tin- and germanium-containing species applied to the production of antiradiation drugs and the curing of pneumosclerosis.[472,473]

III. THERMAL STABILITY OF METAL-CONTAINING POLYMERS

In most cases the thermal stability of metal-containing polymers is much higher than that of their metal-free analogues. It is not uncommon that the decomposition temperatures (T_d) of such macromolecules exceed 300-400°C. Thus, for potassium and sodium methacrylates these values are equal to 420 and 470°C, respectively.[474] Other authors report that the onset of decomposition and the maximum rate of this process for alkaline metal polymethacrylates are the same, 350 and 450°C, respectively.[475]

Interestingly, the enhanced thermal resistance of metal-containing polymers also is often typical of MCM copolymers with conventional monomers. Thus, T_d of MMA-MAA copolymers containing no more than 1 mol% of lithium, sodium, or potassium acrylate are higher by 25, 28, and 31°C, respectively.[476]

The thermal stability of copolymers composed of styrene and lithium or potassium acrylates was compared with that of polystyrene. It was shown to be practically independent of metal content in the copolymer, whereas even small additives of sodium acrylate led to a significantly higher thermal stability.[477] However, styrene copolymers with alkaline-earth metal acrylates were found to be 30 to 50°C less thermally resistant than polystyrenes.[478] A moderate thermal stability is also typical of some Hg(II)-containing polymers: polybis(*p*-vinylphenyl)mercury and phenylmercury polymethacrylate decompose at 240 to 250°C with metal liberation.

Transition metal-containing polymers are also distinguished by a significant thermal stability. Thus, polyoxyethylmethacryloxytributoxytitanium does not undergo destruction up to 400°C.[479] In the vicinity of this temperature Co(II), Ni(II), and Zn(II) polyacrylates are equally decomposable (Table 17).[160] Weight losses at temperatures up to 210°C are caused by polymer-occluded methanol liberation (a solvent used for conducting the polymerization). The second endothermic peak relates to a more vigorous decomposition of polymeric salts. Generally, however, the thermal polyacrylate stability changes in the series Zn(II) > Co(II) > Ni(II) > Cu(II). It is interesting that thermal stability of polymers based on these MCMs and styrene changes in the same series as in the case of homopolymers at 260 to 330°C.[480] As this takes place, the stability of the Zn(II) acrylate copolymers increases with MCM content. In the case of other metal acrylates the effect is quite the reverse. Another example is Zn(II) ethylene glycol methacrylate phthalate which leads to a higher stability of the MMA-hydroxyethyl methacrylate copolymer. However, copolymers based on a similar Cr(III) salt are less stable than the corresponding metal-free analogues.[123,124]

As a matter of fact, the thermostabilization effect can be equally attained by simply mixing metal-containing polymers with conventional high molecular compounds. This can be exemplified by a method with which calcium, barium, iron, and copper polyacrylates are added to curable polyester resins, and polymethacrylates, itaconates, and maleates of various metals or the corresponding copolymers to PVC.[482-484] However, the best results are usually obtained in those situations when metal-containing groups are inserted in the polymer chain. Thus, for example, the thermal stability of a vinyl chloride-Mg(II) acrylate is nearly twice as high as that of a PVC copolymer, and the same effect is observed for Ca(II) and Zn(II) acrylates.[485-487]

A high stability is also characteristic of polymers containing *nV*-type MCMs. Thus, the T_d values of polymers obtained from Co(1-vinyl-2-alkylbenzylimidazole)$_2$Cl$_2$ were 310 to 500°C,[254] and still higher values (400 to 600°C) were observed for Zn(II) adducts.[257]

Sometimes, oxidative destruction of metal-containing polymers takes place at fairly low temperatures. For instance, in the presence of oxygen, sodium polyacrylate is even less stable than the parent MCM (decomposition at 270°C).[65] For titanocene polymethacrylates T_d was 220°C.[96] The thermooxidative destruction of other Ti(IV)-containing polymers also occurs rather readily, the weight losses at 200°C being as large as 60%.[65] At the same time, T_d in the oxidative destruction of Ca(II), Mg(II), and Zn(II) polyacrylates was very high (400 to 500°C).[65] The T_d values for copolymers of potassium methacrylate (1 to 2 mol% of MCM) with MMA and MAA exceeded their MCM-unmodified counterparts.[476] Methacrylates of other alkaline metals were also useful in significantly enhancing the thermooxidative stability of these copolymers.

Table 17 **Data on thermal decomposition of metal polyacrylates under nitrogen[160]**

		DTGA[a]		TGA		DTA	
Polymer	**Process**	**Temperature peak °C**	**Temperature range °C**	**Weight loss %**	**Total weight loss % within 25–500°C**	**Temperature peak °C**	**Temperature range °C**
Zn(II) polyacrylate	CH_3OH losses	84	<210	14.5	51.5	90	<210
	Decomposition	418	360–460	35.0		416	397–440
Co(II) polyacrylate	CH_3OH losses	96	<210	18.0	60.5	110	<210
	Decomposition	410	370–480	39.4		413	403–450
Ni(II) polyacrylate	CH_3OH losses	93	<210	22.0	65.5	102	<210
	Decomposition	368	300–415	40.0		364	360–415

[a]DTGA, differential thermogravimetric analysis.

In most cases metal-containing polymers are also characterized by high T_g values. This is true, for example, of ternary polymers containing alkaline metal acrylates: MMA + MAA (styrene) + MCM.[488,489] In these cases the T_g rise was as high as 8 to 14°C. This is accompanied by a flow temperature (T_f) increase.

Copolymers with other alkaline metal-containing monomers, such as AN with sodium metallyl sulfonate, also showed higher T_g and T_f values.[490] A T_g increase, as compared with PMMA, was also observed for an MMA copolymer with Cr(II) ethylene glycol methacrylate phthalate;[124] as well as MMA with MMA complexes and UO_2(II), Pr(III), and other chlorides (nitrates).[345]

Some exclusions to these relate to lower T_g values for PMMA as compared with those for an MMA-Cu(II) acrylate copolymer (from 105°C for PMMA to 60°C when the copolymer contained 7.1 wt% of Cu(II).[217] A similar effect is produced by tributyltin methacrylate (TBTM).[491]

Of some interest are the thermomechanical properties of Ti(IV)-containing polymers.[479] Unlike their "metal-free" analogues, they have two T_g's and exhibit high-elasticity characteristics, the lower T_g corresponding to hydrogen bond breakage. The thermomechanical curves for polyacrylates of other transition metals, such as Co(II) and Ni(II), are distinguished by a typical deformation-temperature dependence in that the transition from the glassy to the viscous-flow state is characterized by the appearance of a high-elasticity plateau (Figure 32). On the other hand, PAA is not meltable and resists high-elasticity deformation.

The thermomechanical properties of copolymers depend on the content of MCM units. Thus at low contents, 3 to 5 wt% of Co(II) and Ni(II) acrylates in their copolymers with styrene, the properties of such products are very much the same as those of polystyrene.[479] However, as the MCM content is increased to 10 to 12%, the thermomechanical plot starts showing a high-elasticity plateau with a low 2 to 6 wt% degree of deformation. As this takes place, T_g runs 250 to 300°C.

The enhanced thermal stability of metal-containing copolymers is largely due to the cross-linking effect of MCMs containing two or more polymerizable groups. Because the cross-linking proceeds through a metal ion, the resulting network has a lower strength than in the case of polymers cross-linked through covalent bonds. However, as follows from the available data, the effect of such a "soft" cross-linking is still significant. Moreover, it can be supplemented with interchain interactions.

Cross-linked structures are formed during homo- and copolymerization of Mg(II), Sr(II), Ba(II), Ge(IV), Pb(II), Zn(II), Sn(IV), Ti(IV), Zr(IV), Hf(IV), V(V), and Cr(III) methacrylates and itaconates.[96,492-496] Furthermore, cross-linking takes place in the polymerization of Co(II), Ni(II), and Zn(II) acrylates[146,160] with its degree changing as follows:

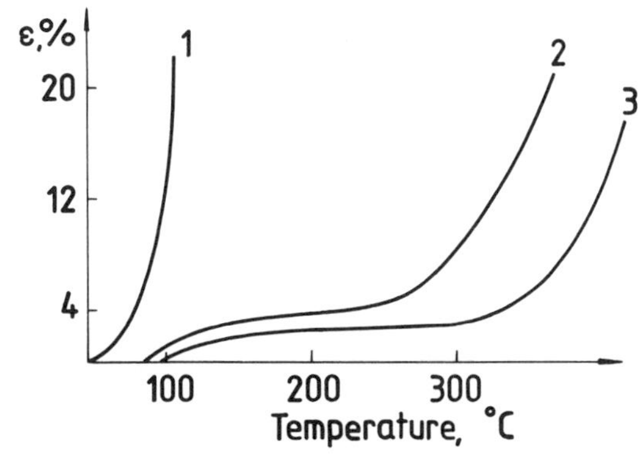

Figure 32 Thermomechanical plot of PAA (1), Ni(II) polyacrylate (2), and Co(II) polyacrylate (3).

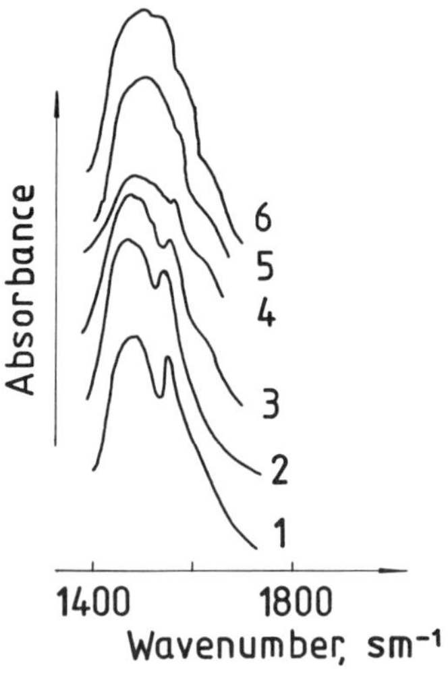

Figure 33 Fragments of IR spectra for Co(II) acrylate (1), polymers based on this acrylate (2), and polymers heated in air for 30 min at 100°C (3), 150°C (4), 200°C (5), and 250°C (6).

Zn(II) > Co(II) > Ni(II), i.e., in the order opposite to changes in the content of unreacted double bonds (9.15 and 27%, respectively).

An alternative mechanism consists in repolymerizing the residual double bonds.[146]

$$—CH_2—\underset{|}{C}H—CO—O—M—O—CO—CH{=}CH_2 \xrightarrow{\text{heating}}$$

$$—CH_2—\underset{|}{C}H—CO—O—M—O—CO—CH—\underset{|}{C}H_2—$$

Thus, in the case of Co(II) polyacrylate heated at enhanced temperatures for 30 min, one can observe a significant drop of intensity in the infrared (IR) spectrum of the $\nu_{C=C}$ band (1640/cm), which disappears completely at 250°C (Figure 33).

The cross-linking effect of divinyl MCM also has been reported for finished polymers, e.g., in the case of $Zn[OCOCH{=}CHCONH_2(CH_2)_6NH_2]{\cdot}4H_2O$ in a mixture with PE or PP.[112]

Note that in addition to the cross-linking effect of MCMs, there may be some other explanations for the larger T_g values of metal-containing polymers (such as the effect of the bulky side groups,[497] a smaller free volume in the polymeric matrix as a result of a specific interaction between MX_n and the polymer,[345] etc.).

IV. RADIATION STABILITY AND PHOTOPHYSICAL PROPERTIES OF METAL-CONTAINING POLYMERS

In determining the effect of various radiations on metal-containing polymers, one of the major factors is constituted by metal content in a given substance. At low concentrations the metal acts as an agent conveying the radiant energy to the polymeric matrix with the result that the degree of its destruction or cross-linking (depending on matrix type) becomes more pronounced. The instability of some metal-containing copolymers incorporating small amounts of metals, predetermines major fields of their applications. Thus MMA copolymers with Pb(II), Ba(II), Ca(II), and Sr(II) are used as electronic components or photoresists displaying high sensitivity and resolving power, as well as film-forming capabilities and thermal stability.[498,499] Generally, the resist sensitivity is proportional to the atomic number of a metal; therefore, the best results are obtained with Pb(II) methacrylate. The sensitivity of its polymer with MMA (a metal content of 3 wt%) to electron beams and X-ray radiation exceeds that of PMMA by a factor of 3. Using metal-containing compositions makes it possible to produce contrast images even in the case of thick films.

On the contrary, a TBTM-allyl chloroacetate copolymer undergoes a radiation-induced cross-linking to give an insoluble product. This permits the use of this copolymer as a photosensitive material.[500]

At high metal contents (30 to 50 mol%), polymers acquire a "protective" ability with respect to various radiations. For example, this ability is typical of Mg(II), Zn(II), Sr(II), Cd(II), Cu(II), Hg(II), and Pb(II) homopolymers; and of their copolymers with MMA and tetraethylene glycol diacrylate.[501] In this case, the protective properties are coupled with an optical transparency of the materials. Thus, in the case of Pb(II) methacrylate copolymers with allyl acrylates, the relative transparency is as high as 88%. This material is used for making clear X-ray compensating filters.[502] Copolymers of lead monoalkyl itaconates are also used as radiation-protective materials.[502] Copolymers of Cd(II) methacrylate with alkyl acrylates are finding applications for neutron-protection purposes.[504]

Interesting processes occur as some metal-containing polymers are exposed to nonionizing radiation. Thus, products resulting from the electrochemical polymerization of Fe(II), Ru(II), and Os(II) VP complexes are photoconductive.[505] In the last few years there have been works dealing with fluorescent properties of Eu(III) polymers containing rare-earth metals, primarily Eu(III) and Tb(III). These properties are much more pronounced than in the case of their monomeric and low molecular analogues. The characteristic emissivity of a ternary copolymer of Eu(III) methacrylate with MAA and MMA exceeds that of Eu(III) octanoate inserted into a PMMA–PMAA matrix at the same Eu(III) content equal to 0.01 wt% by one order of magnitude.[98] A similar effect has been observed for Tb(III): the intensity was higher by a factor of 3. Tercopolymers of Eu(III) and Tb(III) triundecylenates with MMA and MAA[200] are also distinguished by a high fluorescent efficiency.

However, the highest fluorescent efficiency is attained in systems where organic ligands operate as energy donors for MX_n with the maximum coordination number (CN). This is a case when ligands are represented by compounds containing aromatic rings and when metallochelate-type MCMs are applied to polymerization conversions. The fluorescent properties of such polymers prove much better than those of polymer-analogous conversions. Thus, the fluorescent intensity of a copolymer composed of Eu(III) 2-(4′-vinylbenzoyl benzoate)di(2-benzoyl benzoate) and styrene:

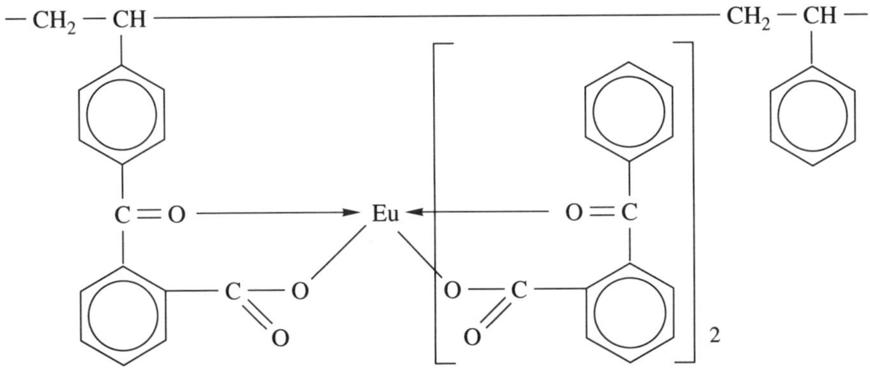

at a metal content over 1 wt% is much higher than that of the immobilized $EuCl_3$ products on a partially substituted 2-carboxylbenzoyl- or 2-carboxyl-naphtoyl polystyrene.[386] The effect of the ligand nature on this phenomenon has been exemplified by MMA copolymers with 4-methyl-4′-vinylpyridine complexes:[381]

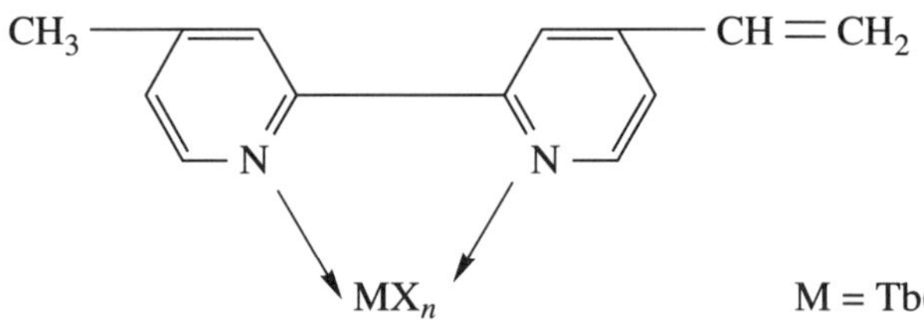

M = Tb(III) or Eu(III)

For Tb(III) and Eu(III) complexes the following order of changes in the fluorescent intensity was observed: 2, 2′, 2′′,-tripyridyl (Tpy) > Dipy > Phen and dibenzoylmethane > 1-furyl-4-trifluoroacetylacetone > Dipy, respectively. It is interesting that the fluorescent intensity in the copolymers was far higher than that in the parent MCM, suggesting that fluorescence can be extinguished by the double bond.

Other metal-containing polymers based on $[Ru(Dipy)_2(4\text{–}VP)_2]^{2+}$ and $[Ru(Dipy)_2(4\text{–}VP)Cl]^{+}$ in methanol or dilute acids at enhanced temperatures are also capable of fluorescing.[371] Thus, in methanol at 455 nm the excited metal ions in $[Ru(Dipy)_2(4\text{–}VP_2)]^{2+}$ were found to fluoresce within a broad band with a maximum at 615 nm; whereas in the case of poly$[Ru(Dipy)_2(4\text{–}VP)Cl]^{+}$, the excitation at 505 nm gave rise to emission at 708 nm, the emission spectra of these polymers at –196°C being similar to those obtained at room temperature with a far lower intensity.

Some metal-containing polymers are characterized by an important photophysical property consisting of a selective light absorption. Thus in Nd(III) acrylate copolymers with MMA, styrene, and other comonomers, light absorption typical of this metal takes place at 580 nm.[97] When such copolymers are used as light filters for metal-halide and mercury lamps, the light absorbed is superimposed on the emission spectra with the result that glare is suppressed.

Another optical effect resulting from metal addition to polymers demonstrates itself in tercopolymers composed of MMA; MAA; and barium, lead, sodium, or strontium methacrylates.[492] These products are distinguished by enhanced integral light transmission values as compared with those typical of their "metal-free" analogues. At 250 to 700 nm copolymers composed of MMA, MAA, and alkaline metal methacrylates display a high spectral transparency exceeding that of PMMA.[506] The highest transparency in the visible spectral region is exhibited by copolymers synthesized in conditions when the content of lithium and sodium acrylates in the starting mixture is equal to 0.10 to 0.50 mol%. The spectral transmittance of these acrylates at 400 to 700 nm is as high as 90.5 to 97.0% against 86.0 to 91.0% without MCMs and 87.5 to 93.5% for PMMA.

An interesting feature of transition metal-containing polymers is that they change color. This change can be associated with the residual unsaturation in the polymer. Thus, a pink reaction mixture composed of MMA, MAA, and Co(II) methacrylate turns green when the number of the C=C bonds decreases by 15%.[185] Copolymers of Cr(III) alkylvinyl sulfoxide complexes with MMA display thermochromatic properties in that they change from emerald-green (20 to 30°C) to violet (40°C).[507] It is obvious that changes in the configuration of the complex under the effect of temperature is due to the presence of a polymeric matrix with the rigidly immobilized complex:

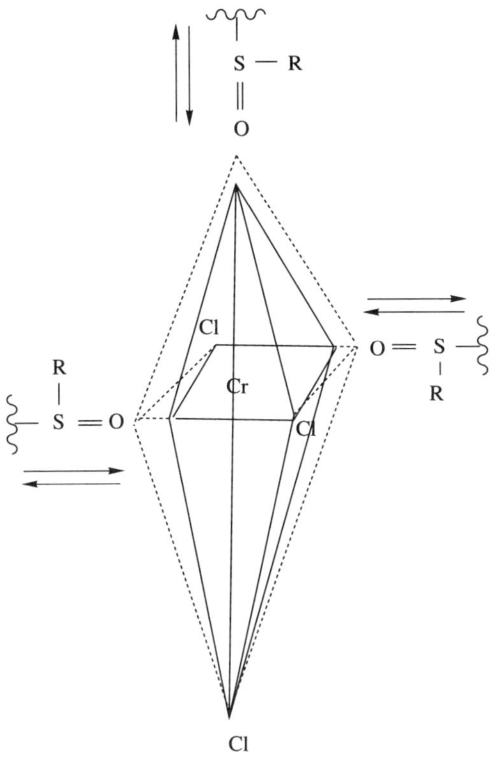

As the temperature decreases, the matrix rigidity increases, leading distortion of the macrocomplex configuration. On the other hand, with increasing temperature, the polymer chain mobility grows, the influence of the matrix becomes weaker, and the complex takes the original configuration. Note that addition of chemically loose Cr(III) dialkyl sulfoxide complexes did not lead to thermochromatic effects.

Also, polymeric products can be colored by using other MCMs with a more complicated structure. Thus, a stable and uniform color of polyolefins can be attained with the aid of a polymerization catalyst in the form of a product resulting from the interaction of Ti(IV), V(IV), and V(V) with diazoacetylene alcohols containing—along with a multiple bond—a chromophore group.[508]

V. ELECTROPHYSICAL PROPERTIES OF METAL-CONTAINING POLYMERS

A metal ion present in the polymer can be responsible for its electroconductivity. However, nearly all metal-containing polymers under discussion are insulators in themselves. The metal conduction zone in such systems is obviously located at a very high energy level. At the same time, conductivity can be enhanced by increasing either temperature or metal content in the polymer. Thus, the Cu(II) conductivity can vary widely, from 10^{-11} to 10^{-6}/Ohm/cm.[217] At 90°C the conductivity of the polyacrylate of another metal, namely, Cr(III), varies as a function of metal content within $6.0 \cdot 10^{-7}$ to $2.4 \cdot 10^{-9}$ Ohm/cm.[86] It is evident that in some cases these metal-containing polymers can behave as semiconductors. Still higher conductivity is displayed by products resulting from the polymerization of AAm complexes of Er(III), Ca(II), Cr(III), and other metal nitrates at 20 to 25°C (10^{-5} to 10^{-7}) ohm/cm.[218]

Another property typical of some polymers such as a TBTM–MMA copolymer (an MCM content of 25 to 30 mol%) is piezoelectric activity, i.e., an electric polarization resulting from metal contraction or expansion in specified conditions.[491] In the unpolarized state piezoelectric properties are only exhibited by films of copolymers obtained from cyclohexanone and methyl ethyl ketone. This might be due to the fact that such solvents cause the orientation of the MCM dipoles due to hydrogen bonding. It is assumed that the whole of the solvent is removed from the polymer, so that the orientation of the dipoles at the time of film-formation and drying is easier to maintain. This leads to anisotropy responsible for the piezoelectric activity.

Note also that polymeric VP complexes of Fe(II), Ru(II), and Os(II) possess electroactive properties.[372] Coatings prepared from these materials are distinguished by the fact that the redox sites are contained in all polymer units. The redox conductivity permits the polymeric film to grow continuously. In this case, it can behave as an electron-transferring agent at the film-solution interface.

Finally, particular emphasis should be placed on the fact that metal-containing copolymers are ideal substrates for a subsequent synthesis of superconducting ceramics. In fact, such polymers alone provide for a truly uniform distribution of metal ions to eventually ensure good performance of the ceramic materials obtained. Thus, a Y(III)-ceramic material produced from a copolymer of Y(III), Ba(II), and Cu(II) acrylates exhibits a "sharp" superconducting junction at 94 K without additional annealing under oxygen; and a Bi(III) material synthesized from the products of the spontaneous polymerization of AAm complexes composed of Bi(III), Pb(II), Ca(II), Sr(II), and Cu(II) nitrates represents a stable single-phase system with a junction temperature of 110 K.[325,507] In both cases the volume of the diamagnetically screened phase is responsible for 100%.

VI. METAL-CONTAINING POLYMERS AS ELECTROLYTES

Ionic-type metal-containing polymers comprise a peculiar kind of cross-linked compounds, on the one hand, and polyelectrolyte with the associated charges interactions, on the other hand. The existence of a significant number of ionogenic groups in the polymers is responsible for the fact that in some conditions they swell, which distinguishes them from their metal-free analogues. Therefore, copolymers containing alkaline metal acrylates exhibit a remarkably high ability to absorb water.

Homopolymers of the above acrylates[509-511] and more commonly their copolymers with AA, MAA, AAm, etc.[512-517] are used as water-absorbing materials. One more way of producing polymers with such properties is graft copolymerization of a mixture containing alkaline metal and AA acrylate onto cellulose, starch, and other polysaccharides.[518]

Water absorption is maximum for modified potassium 2,2′-azobis(2-amidopropane) (Table 18).[511] This is coupled with one more valuable property, i.e., a high initial water absorption rate. Note that the resulting polymer gels, e.g., sodium polyacrylate at 170 to 230°C, can be dried. In other words, water-absorbing polymers can be used repeatedly.

This property of alkaline metal acrylate copolymers underlies their uses, such as various absorbents of water,[517] blood, [520] and urea;[521] swaddling clothes, bandages, diapers, and medical tampons;[514,518] evaporation retarders in water reservoirs;[516] water-isolating agents in oil production;[522] drilling mud water-removal compounds,[523] etc. The polyelectrolyte properties of ionic-type metal-containing polymers ensure the application of these materials to flocculation and coagulation processes. In this field, the best results are achieved through the use of sodium and magnesium methacrylate copolymers, as well as a graft copolymer of sodium acrylate on polyvinyl alcohol.[524-533]

Table 18 **Water absorption with some polymers based on alkaline metal acrylates**

Polymer	Water absorption, g/g polymer: From water	From 0.9–1.0 % NaCl solution	From urea solution	Ref.
Sodium polyacrylate	870	70	–	509
Modified potassium polyacrylate	6200 (10 s) 9600 (15 min)	730 (10 s) 970 (15 min)	– –	511
Graft copolymer of sodium acrylate with AA on a starch	615	91	75	518

VII. METAL-CONTAINING POLYMERS AS INGREDIENTS OF POLYMER COMPOSITIONS

Metal-containing polymers that comprise physically homogeneous systems as opposed to metal-filled compounds, show good results when used as ingredients of various polymer compositions. Among the important uses is finishing of inorganic fillers, e.g., silica gels, carbon, and chalk for ensuring their compatibility with hydrophobic polymers. Thus, adsorption of an MAA-sodium methacrylate copolymer (21 wt% of MCM) onto a technical-grade carbon surface enables one to obtain fillers with a specified amount of oxygen-containing functional groups.[534]

Another finishing method consists of MCM polymerization directly on a filler surface, particularly, radical polymerization of Ti(IV), Co(II), and Ni(II) acrylates on a silica gel surface.[479] The treated filler contained 0.5 to 2.0 wt% of adsorbed metal-containing polymer.

This operation also can be accomplished via a "chemical" MCM grafting on a filler surface:

$$\begin{aligned}&SiO_2-OH+Ti(OR)_3(OXCH=CH_2)\rightleftarrows\\&SiO_2-O-Ti(OR)_2(OXCH=CH_2)+ROH.\end{aligned}$$

The resulting immobilized MCM can enter into a copolymerization reaction with the result that the product obtained will contain a chemical bond between the polymer and the filler. The finishing operation gives filler-PE mixtures with significantly improved physicomechanical properties.

One more way for producing metallopolymer compositions is by polymerization filling. It is based essentially on monomer polymerization in the presence of a dispersed filler with a surface-applied catalytic system component. Also, this can be done using Ti(IV)-containing metallopolymer catalysts. Such a method was instrumental in producing polymer-metallopolymer compositions including PE-polydimethylvinylethynylcarbinoloxytributoxytanium directly during ethylene polymerization.[479] On being added to a SiO_2–PE system (30:70 by weight), 0.5 wt% of this metallopolymer gave significantly improved physicomechanical properties of the product.

In addition, metal-containing polymers can be used as adhesive additives. To do so, use is made of polymers containing both nontransition and transition metals. Thus, Al(III) polymethacrylate exhibits improved compatibility with hydrophobic materials.[535,536]

Trialkyltin acrylate polymers adhere perfectly to glass and metals.[537] Also, a fairly high adhesion is provided by metal-containing compositions. Thus, addition of 0.5 and 2 wt% of nickel polyacrylate to PE allowed it to enhance its steel adhesion by a factor of ca. 7 and 13, respectively.[479]

It is interesting that grafted metallochelate polymers are used as antiwear and antitear additives to plastic lubricants.[538] In this case, the grafted polymer was found to be more efficient than unchelated Cu(II) polyacrylate.[539]

VIII. SELECTIVE ADSORPTION AND METAL ION BINDING

Some metal-containing copolymer systems are of interest for producing selective sorbents for metal ions. The effect of such sorbents is based on the so-called "template" effect, i.e., "recognizing" and binding those metal ions that are used as "blanks" in the synthesis of metallopolymer complexes.[540-543] Selective sorbents are synthesized via polymerization of coordination- or chelate-type MCMs using cross-linking agents. In so doing, ion removal leads to a significant growth of the sorption capacity and rate, as well as selectivity of the "tuned" polymers. The synthesis of such polymers can be described as follows:

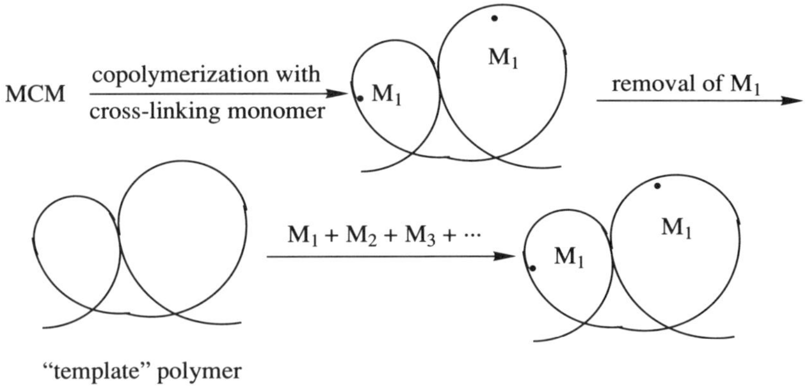

In this case the template polymer will have "pockets" specially fitted to just those metal ions that were removed from the polymeric matrix.

Thus such ternary polymers as $M(VIA)_4(OCOCH_3)_2$—where M = Co(II), Ni(II), and Zn(II) with *N*-vinylpyrrolidone (VPr) and divinylbenzene—were used for this purpose.[342] The copolymers were crushed, and the metal ions employed as the template were washed off by keeping the powder in 0.1 *N* HCl.

Each template copolymer is capable of sorbing any metal ion better than the control copolymer does. As this takes place, the copolymer tuned to Co(II) acts as a good selective sorbent of these ions, whereas no selective sorption was observed with respect to Ni(II) and Zn(II).

Better results were achieved when metallochelate monomers were used for producing selective sorbents. Thus, a copolymer of 4-vinyl-4′-methyl-2,2′-dipyridyl, styrene, and divinylbenzene from salt solutions with the same concentrations of Ni(II) and Cu(II) provides a better binding of Cu(II) due to a higher binding constant for this ion; in contrast, the same copolymer formed in the presence of Ni(II) displays the template effect with respect to the latter.[544]

Polymer chelates obtained by copolymerization of dithiophosphate complexes (e.g., bis-di(4-vinylphenyl)dithiophosphinate of Ni(II), or Co(II) or Cr(III) with styrene and ethylene glycol methacrylate, after they have been freed of metals) proved selective with respect to their own ions. For Ni(II) and Co(III) the ratios of the amount of the metal bound to the tuned and "untuned" polymers, were 1.4 and 2.4, respectively.[545]

New types of tuned polymers, a model of the histidine moiety, have been synthesized.[546] Complexes composed of Co(II) and 4(5)-vinylimidazoles and leucine-2-picolinidine or leucine are copolymerized with divinylbenzene in a water-alcohol-butanol mixture, the polymer particles sized 200 to 500 nm are extracted with methanol, and the template molecules are removed with Dipy. The resulting tuned polymer is used as a catalyst for hydrolyzing *p*-nitrophenyl esters to aminoacids:

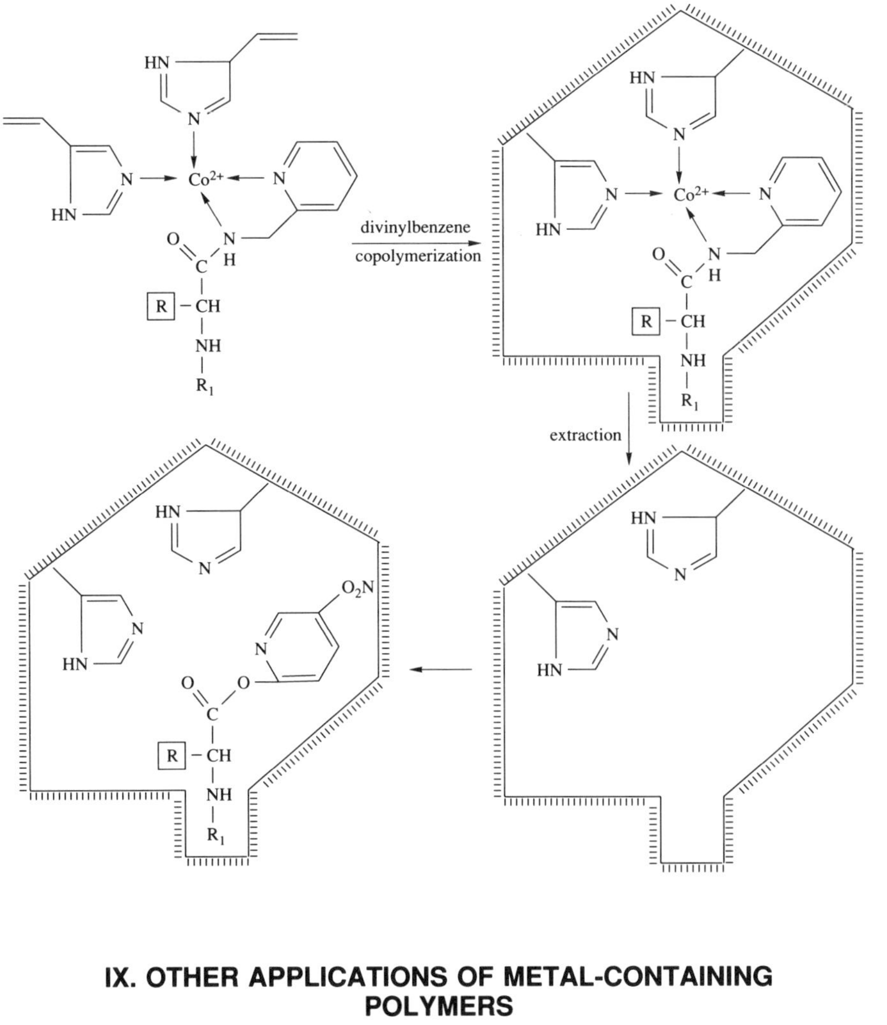

IX. OTHER APPLICATIONS OF METAL-CONTAINING POLYMERS

The above data do not exhaust the great diversity of properties and applications of metal-containing polymers. It is noteworthy, for instance, that composites containing nontransition metal acrylates (sodium, potassium, zinc, magnesium, calcium, and aluminum), as well as sodium acrylate copolymers grafted onto PVC display high fire-resistant properties.[547-549] Note also the application of methacrylates of Cu(II), Mg(II), and alkaline-earth and alkaline metals, to name a few, as components of soil structuring additives.[550-552] In this case, the cross-linking effect of MCMs is usually complemented with the action of conventional cross-linking agents.

Of great interest are applications of MCM polymerization in the atomic industries. Thus formation of a copolymer including a complex of an UO_2(II) and VIA or its methyl derivative, on the one hand, and excess ligand, on the other, was proposed to be used for nuclear waste reclamation.[343] In this case, the chemical binding suppresses the leaching of radioactive materials and thereby improves the environmental situation. One more application in this area is the effective separation of ^{235}U and ^{238}U stated above, relying on different polymerization rates of acrylates of these isotopes in a magnetic field.[101]

In summary it is evident that metal-containing polymers are distinguished by a great variety of properties that predetermine their applications. At the same time, most materials such as these display a high heat resistance. On the other hand, other properties and uses are largely very unique and typical of specific polymer species. Thus, water-absorbing compositions are exclusively composed of alkaline metal copolymers, whereas structurizing agents include alkaline and alkaline-earth methacrylates. Polymethacrylates of various metals are mostly used for protection against various radiations. Polymers containing π- and ionic-type MCMs possess biocidic activity, and donor-acceptor species are efficient catalysts of a number of processes for producing metal ion sorbents. Polymers obtained from chelate-type MCMs based on rare-earth metals exhibit good fluorescent characteristics.

In the future, the emergence of novel applications for these unique polymers should be expected.

Conclusion

A search for nonconventional tools of synthesis, as well as for accessible methods of MCM polymerization, such as may be expected from the hands of a narrow group of highly proficient specialists, may come under control of any researcher. This is the way a breakthrough in this field will be accomplished. It will require the application of joint efforts of scientists who have interests in a variety of areas and who will be satisfied by the ensuing benefits. Among them is the possibility of deriving new valuable information and unusual properties of the resulting metal-containing polymers. Even now the existing level of knowledge permits prediction of composition and structure of products that would best satisfy our practical needs.

At the same time, as in any other developing branch of science, MCM chemistry is characterized by the existence of problems that remain to be solved. First of all, this relates to the development of production processes for the synthesis of new MCM classes such as chelate, cluster, and heterometallic types. Significant progress is expected in the following areas of polymerization conversions: synthesis of polymers composed of different units including transition metals in a number of valent states and having dissimilar metallocenter geometries. There is every reason to believe that in the near future we shall witness the synthesis and copolymerization of MCMs incorporating optically active groups that are chemically bonded to metal ions; the same is true of MCMs with chromophorous, stabilizing, biologically active fragments, etc. As a matter of fact, the mutual copolymerization of MCMs themselves has been little studied, if at all; the resulting heterometallic polymers may be expected to possess unusual electroconductive, photochemical, and other properties. A very interesting aspect of these works is the "assemblage" of MCMs with participation of surface metals in that their polymerization would lead to the production of thin layers on metal surfaces for the purpose of corrosion inhibition.

It is reasonable to expect considerable advances in the application of metallopolymers to silver-free photographic processes, to those associated with the recording, storage, and treatment of information; and to holography, optoelectronics, laser technology, etc. Perceptible results have been obtained in the use of metal-containing polymers in studies of carcinogenesis, photosynthesis, and enzymatic catalysis, to name a few. It is very important that the properties specific to metallopolymers be combined with those typical of polymers as a whole (such as a filming ability, thermoresistance, and electric and mechanical strength).

It would be no exaggeration to say that at the present time metallopolymers have come into their own in various fields of catalysis as immobilized metallocomplex compounds.[1] These catalytic requirements, to a large measure, have prompted a search for and the development of new types of metallopolymers.

Yet polymerization conversions of MCMs represent no more than one of the ways for accomplishing chemical syntheses with their participation. In the last few years, there has been application of MCMs as substrates in a host of reactions (such as transition metal acrylate hydrogenation[553] and hydroxylation of titanates with alkyl aromatic radicals[554]) and in interactions of titanium and zirconium alcoholates with benzyl halides and carbon monoxide to produce the corresponding esters,[555] etc. Apparently, this group includes MCM epoxidation, ozonization, amination, and other processes to give remarkable products.

Finally, special mention should be made that the ideas under discussion for which appearance and development have been generated by a need for the design and polymerization of MCMs are not limited to MCMs as such. Thus, many problems in the chemistry of naturally occurring compounds are being solved using homo- and copolymerization approaches, condensation of vinyl derivatives of pyridine and purine-type nucleic acids, unsaturated derivatives of alkaloids, acrylic and methacrylic esters of cholesterol, and steroid hormones. Eventually more will undoubtedly be added to this list.

REFERENCES

1. Pomogailo, A. D. (1988). *Polymeric Immobilized Metallo-Complex Catalysts.* Nauka, Moscow.
2. Pomogailo, A. D. and Savost'yanov, V. S. (1985). *J. Macromol. Sci., [Part C] Rev. Macromol. Chem., Phys.,* 25:375.
3. Pomogailo, A. D. and Savost'yanov, V. S. (1988). *Metal-Containing Monomers and the Associated Polymers*, Khimiya, Moscow.
4. Pittman, C. U., Jr. (1977). In *Organometallic Reactions*, Becker, E. I. and Tsutsuit, H. Eds. New York, 6:1.
5. Carraher, C. E., Jr., Sheats, J. E., and Pittman, C. U., Jr. (1978). *Organometallic Polymers*, Academic Press, New York.
6. Carraher, C. E., Jr., Sheats, J. E., and Pittman, C.U., Jr., Eds. (1981). *J. Macromol. Sci., [Part A] Chem.*, 16:1.
7. Sheats, J. E., Carraher, C. E., and Pittman, C. U., Jr., Eds. (1985). *Metal-Containing Polymer Systems*, Plenum Press, New York.
8. Normant, H. (1959) *Bull. Soc. Chim. Fr.,* 1764.
9. Kochkin, D. A. and Azerbayev, I. N. (1968). *Tin- and Lead-Containing Monomers and Polymers,* Nauka, Alma-Ata.
10. Joffe, S. T. (1958). *Usp. Khim.*, 27:1010.
11. Hatada, K., Nakanishi, H., Ute, K., and Kitayama, T. (1986). *Polym. J.*, 18:581.
12. Henry, M. C. and Noltes, J. C. (1960). *J. Am. Chem. Soc.,* 82:555.
13. Rosenberg, S. D., Gibbons, A. J., Jr., and Ramsden, H. E. (1957). *J. Am. Chem. Soc.,* 79:2137.
14. Langer, H. G. (1961). *Inorg. Nucl. Chem.*, 21:297.
15. Labadie, J. W., McDonald, S. A., and Willson, C. G. (1986). *Am. Chem. Soc. Polym. Prepr.*, 27:216.
16. Nesmeyanov, A. N., Borisov, A. E., Savel'yeva, I. S., and Golubeva, E. I. (1958). *Izv. Akad. Nauk SSSR OKhN*, 1490.
17. Bartocha, B., Bilbo, A. J., Bublitz, D. E., and Gray, M. Y. (1961). *Z. Naturforsch., Teil B,* 16:357.
18. Juenge, E. C. and Cook, S. E. (1959). *J. Am. Chem. Soc.*, 81:3578.
19. Krongauz, E. S. (1966). In *Advances in the Field of Organoelement Polymer Synthesis*, Korshak, V. V., Ed. Nauka, Moscow, 129.
20. Braun, D. (1961). *Angew. Chem.,* 73:197.
21. Koton, M. M., Kiseleva, T. M., and Florinsky, F. S. (1959). *Izv. Akad. Nauk SSSR OKhN*, 948.
22. Bachman, B., Carlson, C. L., and Robinson, M. (1954). *J. Am. Chem. Soc.*, 73:1964.
23. Korshak, V. V., Polyakova, A. M., and Tambovtseva, E. S. (1959). *Izv. Akad. Nauk SSSR OKhN*, 742.
24. Steinbach, M. and Wegner, C. (1977). *Makromol. Chem.*, 178:1671.
25. Fujita, N. and Sonogashira, K. (1974). *J. Polym. Sci.*, 12:2845.
26. Marina, N. G., Duvakina, N. V., Monakov, Yu. B., Kuchin, A. V., and Tolstikov, G. A. (1984). *Dokl. Akad. Nauk SSSR*, 276:635.
27. Feld, R. and Cowe, P. L. (1955). *The Organic Chemistry of Titanium*, Laporte Titanium Ltd., Butterworth, London.
28. U.S. Patent (1954) 2,655,523.
29. U.S. Patent (1954) 2,654,770.
30. Yatluk, Yu. G. and Suvorov, A. A. (1985). *Zh. Prikl. Khim. (Leningrad),* 58:1624.
31. Korshak, V. V., Sladkov, A. M., Luneva, L. K., and Girshovich, A. S. (1963). *Vysokomol. Soedin.*, 5:1284.
32. Suvorov, A. L. and Spassky, S. S. (1959). *Usp. Khim.*, 28:1267.

33. Dzhardimaliyeva, G. I., Pomogailo, A. D., and Shupik, A. N. (1985). *Izv. Akad. Nauk SSSR Khim.*, 451.
34. U.S. Patent (1956) 2,708,205; *Chem. Abstr.* (1956). 50:4211.
35. Pomogailo, A. D., Lisitskaya, A. P., Natsibullin, F. Ya., and Dyachkovsky, F. S. (1976); U.S.S.R. Patent 553,907; *Bull Izobr.* (1981). 34:315.
36. Pomogailo, A. D., Golubeva, N. D., and Kitaigorodsky, A. N. (1991). *Izv. Akad. Nauk SSSR Khim.*, 482.
37. Korshak, V. V., Polyakova, A. M., Mironov, V. F., and Petrov, A. D. (1959). *Izv. Akad. Nauk SSSR OKhN,* 178.
38. Korshak, V. V., Polyakova, A. M., Petrov, A. D., and Mironov, V. F. (1957). *Dokl. Akad. Nauk SSSR*, 112:436.
39. Mazerolles, P. and Lesbre, M. (1959). *C.R.*, 248:2018.
40. Plate, N. A., Maltsev, V. V., Davydova, S. L., and Kargin, V. A. (1966). *Vysokomol. Soedin*, 11:1890.
41. Brinkman, F. E. and Stone, F. G. A. (1959). *J. Inorg. Nucl. Chem.*, 11:24.
42. Longi, P., Bernardini, F., and Colombo, L. (1959). *Rend. Ist. Lomb. Sci. Pt. 1 Classe Sci. Mat. Nat.*, 93A:134.
43. Noltes, J. G., Budding, N. A., and Van der Kerk, G. S. (1960). *Recueil*, 79:408, 1076.
44. Kolesnikov, G. S., Davydova, S. L., Ermolayeva, T. N., Shilova, N. D., and Bykhovskaya, M. D. (1960). *Vysokomol. Soedin.*, 2:567.
45. Dzhardimaliyeda, G. I., Tonoyan, A. D., Pomogailo, A. D., and Davtyan, S. P. (1987). *Izv. Akad. Nauk SSSR Khim.*, 1744.
46. Bravaya, N. M. and Pomogailo, A. D. (1992). *Izv. Akad. Nauk SSSR Khim.*, 1746.
47. Pomogailo, A. D. and Golubeva, N. D. (In press). *Izv. Akad. Nauk SSSR Khim.*
48. Minoura, Y., Suzuki, J., Sakanaka, J., and Dei, H. (1966). *J. Polym. Sci. Part A-1*, 4:2757.
49. Koton, M. M. and Dokukina, L. F. (1964). *Vysokomol. Soedin.*, 6:1791.
50. Korshak, V. V., Polyakova, A. M., and Suchkova, M. D. (1960). *Vysokomol. Soedin.*, 2:13.
51. U.S. Patent (1941) 2,258,718; *Chem. Abstr.* (1942). 36:595.
52. Nesmeyanov, A. N. and Nogina, O. V. (1954). *Izv. Akad. Nauk SSSR Khim.*, 41.
53. Bochkin, A. M. and Pomogailo, A. D. (In press). *Izv. Akad. Nauk SSSR Khim.*
54. Poraj-Koshits, M. A. (1981). In *Crystallochemistry. Advances in Science and Technology*, VINITI, Moscow, 15:3.
55. Zurakowska-Orszagh, J., Skupinska, J., and Suchan, F. (1985). *Polimery*, 30:185.
56. Morawetz, H. and Rubin, I. D. (1962). *J. Polym. Sci.*, 57:669.
57. Gronowski, A. and Wojtczak, Z. (1989). *Eur. Polym. J.,* 25:241.
58. Plochocka, K. and Wojnarowski, T. J. (1971). *Eur. Polym. J.*, 7:797.
59. Saviotti, P. P. and Gilson, D. F. R. (1976). *J. Phys. Chem.*, 80:1057; (1978). 82:1841.
60. Plochocka, K. and Wojnarowski, T. J. (1972). *Eur. Polym. J.*, 8:921.
61. Costaschuk, F. M., Gilson, D. F. R., and St-Pierre, L. E. (1970). *Macromolecules,* 3:393.
62. Costaschuk, F. M., Gilson, D. F. R., and St-Pierre, L. E. (1971). *Macromolecules*, 4:16.
63. Gronowski, A. and Wojtczak, Z. (1989). *Macromol. Chem.*, 190:2063.
64. Beres, J., Wojciechowski, J., Hauke, H., Maciejewski, Z., Hetper, J., and Chmura, G. (1970). Polish Patent 94,753; *Chem. Abstr.* (1979). 90:71770.
65. Balcerowjak, M., Hetper, J., Beres, J., and Plkowska, J. (1977). *J. Therm. Anal.*, 11:101.
66. Hopkins, R. P. (1965). *Ind. Eng. Chem.,* 47:2258.
67. Watine, F. B. and St-Pierre, L. E. (1973). *J. Polym. Sci. Polym. Symp.*, 42:123.
68. Costaschuk, F. M., Gilson, D. F. R., and St-Pierre, L. E. (1970). *J. Phys. Chem.*, 74:2035.
69. Lando, J. B. and Morawetz, H. (1963). *J. Polym. Sci.*, 4C Pt. 2:789.
70. Bowden, M. J. and O'Donnell, J. H. (1969). *J. Phys. Chem.*, 73:2871.
71. O'Donnell, J. H., Pomery, P. J., and Sothman, K. D. (1980). *Makromol. Chem.*, 181:409.
72. Kennard, C. H. L., Smith, G., Greany, T. M., and White, A. H. (1976). *J. Chem. Soc. Perkin Trans.*, 2, 3:302.

73. Spewak, L. L., Iwantchev, S. S., and Iwantcheva, N. I. (1982). *Plaste Kautsch.*, 29:12.
74. Iwantchev, S. S., Spewak, L. L., Khaikin, S. Ya., and Iwantscheva, N. I. (1983). *Dokl. Akad. Nauk SSSR*, 273:124.
75. Spewak, L. L., Iwantchev, S. S., Iwantcheva, N. I., Khaikin, S. A., Smol'yanova, O. V., and Torpov, S. L. (1980). *Kinet. Katal.*, 31:43.
76. Landoll, L. M. and Breslow, D. S. (1989). *J. Polym. Sci., Polym. Chem.*, 27:2189.
77. Koton, M. M., Kiseleva, T. M., and Florinsky, F. S. (1960). *Vysokomol. Soedin.*, 2:1639.
78. Leites, L. A., Egorov, Yu. P., Kolesnikov, G. S., and Davydova, S. L. (1961). *Izv. Akad. Nauk SSSR OKhN.*, 1976.
79. Mueller, M., Besecke, S., Schroeder, G., and Arndt, P. J. (1984). German Patent 3,305,121; *Chem. Abstr.* (1984). 101:230775q.
80. Rzayev, Z. M. and Bryksina, L. V. (1974). *Vysokomol. Soedin., Ser. A*, 16:1691.
81. Shaaban, A. F., Azab, M. M., and Messiha, M. N. (1989). *Angew. Makromol. Chem.*, 169:59.
82. Besecke, S., Schroeder, G., Ude, W., and Baumgartner, E. (1984). German Patent 3,224,927; *Chem. Abstr.* (1984). 100:104028q.
83. Wojtczak, Z. and Gronowski, A. (1984). *Polimery*, 27:471.
84. Dzhardimaliyeva, G. I., Pomogailo, A. D., Ponomarev, V. I., Atovmyan, L. O., Shul'ga, Yu. M., and Starikov, A. G. (1988). *Izv. AN SSSR Khim.*, 1525.
85. Selenova, B. S., Dzhardimaliyeva, G. I., Baishiganov, G. E., Efimov, O. N., and Pomogailo, A. D. (1989). *Izv. Akad. Nauk SSSR Khim.*, 1025.
86. Sayyah, C. M., Bahgat, A. A., Sabry, A. I., Said, F. I. A., and El-Hamouly, S. H. (1988). *Acta Polym.*, 39:399.
87. Gonyukh, E. A., Serova, V. A., Khazryatova, L. Kh., Busygina, A. D., and Kuznetsov, E. V. (1982). *Dep. ONIITEKHIM*, No. 890khp-D82; *Ref. Zh. Khim.*, (1983). IB197.
88. Osada, Y. (1975). *Makromol. Chem.* 176:1893.
89. Klyuyev, M. V., Tereshko, L. V., Dzhardimaliyeva, G. I., and Pomogailo, A. D. (1986). *Izv. Akad. Nauk SSSR Khim.*, 2531.
90. Herman, D. F. (1960). U.S. Patent 2,940,957; *Chem. Abstr.* (1960). 54:17960c.
91. Suvorov, A. L. and Spassky, S. S. (1966). *Proc. Inst. Chem. Urals Branch, USSR Akad. Sci.*, 13:39.
92. Suvorov, A. L. and Spassky, S. S. (1964). *Dokl. Akad. Nauk SSSR*, 157:639.
93. Semerneva, G. A., Suvorov, A. L., and Spassky, S. S. (1966). *Proc. Inst. Chem. Urals Branch, USSR Akad. Sci.*, 13:49.
94. Korshak, V. V., Sladkov, A. M., Luneva, L. K., and Girshovich, A. S. (1963). *Vysokomol. Soedin.*, 5:1284.
95. Nesmeyanov, A. N., Nogina, O. V., Berlin, A. M., Girshovich, A. S., and Shatalov, G. V. (1961). *Izv. Akad. Nauk SSSR OKhN*, 2146.
96. Ralea, R., Ungurenasu, G., and Maxim, I. (1967). *Rev. Roum. Chim.*, 12:523.
97. Kozo, I. (1984). U.S. Patent 4,471,078; *Chem. Abstr.* (1984). 100:175928u.
98. Xu, W.-Y., Wang, Y.-S., Zhen, D.-G., and Xia S.-L. (1988). *J. Macromol. Sci.-Chem.*, 27A:1397.
99. Pomogailo, A. D., Selenova, B. S., and Dzhardimaliyeva, G. I. (In press). *Izv. Akad. Nauk SSSR*.
100. Leonov, M. R. and Katkova, M. A. (1989). In Paper Abstr. All-Union Conf. Radical Polymerization, Gorky, 101.
101. Bennett, D. A. (1988). British Patent 2,201,828; *Chem. Abstr.* (1989). 111:30095n.
102. Xu, W.-Y., Pei, Z.-H., Zheng, D.-G., Lou, L.-R., Gao, C.-Y., and Wang, Y.-S. (1990). *J. Macromol. Sci. Chem.*, 27A:1109.
103. Egorov, V. B. and Zubov, V. P. (1987). *Usp. Khim.*, 56:2076.
104. Chen, W. Y., Xu, B. Z., and Feng, X. D. (1982). *J. Polym. Sci., Polym. Chem. Ed.*, 20:547.
105. Chen, W. Y., Sun, D. H., Chang, G. L., Feng, X. D., and Chang, G. B. (1983). *J. Polym. Sci., Polym. Lett. Ed.*, 21:335.

106. Koton, M. M. and Kiseleva, T. P. (1961). *Izv. Akad. Nauk SSSR OKhN,* 1783.
107. Kajzar, F. and Messier, J. (1983). *Thin Solid Films*, 99:109.
108. Japanese Patent (1984). 5,964,615; *Chem. Abstr.* (1984). 101:91678v.
109. Maury, L., Bardet, L., Brunel, R., and Vierne, R. (1981). J. *Raman Spectrosc.*, 11:14.
110. Vansco-Szmerczani, I. and Kallo, A. (1982). *J. Polym. Sci., Polym. Chem. Ed.*, 20:639.
111. Antsyshkina, A. S., Poraj-Koshits, M. A., and Guseinov, M. G. (1974). *Izv. Akad. Nauk SSSR Khim.*, 237.
112. Rustamova, Zh. O., Guseinov, M. G., Ibragimova, G. A., and Guseinova, Z. N. (1983). *Plastmassy*, 8:52.
113. Hempel, A., Hull, S. E., Ram, R., and Gupta, M. P. (1979). *Acta Crystallogr.*, 35:2215.
114. Chen, H., Wu, Y., and Jin, Z. (1990). *J. Appl. Polym. Sci.*, 41:427.
115. Mamedova, S. G., Rzayev, Z. M., Medyakova, L. V., Rustamov, F. B., and Askerova, N. M. (1987). *Vysokomol. Soedin. Ser. A,* 29:1922.
116. Shaaban, A. F., Mahmoud, A. A., and Messiha, N. N. (1988). *J. Appl. Polym. Sci.*, 36:1191.
117. Shaaban, A. F., Arief, M. N. H., Mahmoud, A. A., and Messiha, N. N. (1987). *J. Appl. Polym. Sci.*, 33:1735.
118. Doppert, K., Sanchez-Delgado, R., Klein, H.-P., and Thewalt, U. (1982). *J. Organomet. Chem.*, 233:205.
119. Demchenko, L. M., Kutepov, D. F., and Volgin, A. D. (1975). *Proc. Mendeleev MKhTI*, 86:103.
120. Vancsone-Szmercsanyi, I. (1981). *Mag. Kem. Foly.*, 87, 6:241.
121. Orlin, N. A. and Petrov, K. I. (1976). *Zh. Prikl. Spektrosk.*, 24:171.
122. Orlin, N. A. (1984). In Abstracts of Papers, 4th Int. Symp. Homogeneous Catalysis, Leningrad, Vol. 3, 167.
123. Matsuda, H. and Miyashi, N. (1973). *J. Appl. Polym. Sci.*, 17:1941.
124. Matsuda, H. and Dohi, N. (1973). *J. Appl. Polym. Sci.*, 17:2289.
125. Felicio, E. C. A., De Montauzon, D., and Poliblanc, R. (1984). *Rev. Chim. Miner.* 21:114.
126. Kabanov, V. A. and Topchiyev, D. A. (1975). *Polymerization of Ionizable Monomers*, Nauka, Moscow.
127. Loginova, T. F., Shubin, A. A., Vyalkov, V. V., and Kisel'nikov, V. N. (1983). *Izv. Vuzov, Khim. Khim. Tekhnol.*, 26:1476.
128. Swiss Patent (1979). 609,070; *Chem. Abstr.* (1979). 90:152849e.
129. Beres, J., Zawadski, M., Olkowska, J., Miszka, J., Luszak, M., and Chmura, G. (1979). Polish Patent, 105,454; *Chem Abstr.* (1980). 92:216036p.
130. Larrabee, T. E., Spragce, and Estel, D. (1979). *J. Polym. Sci., Polym. Lett. Ed.*, 17:749.
131. Rindo, K., Yasuda, J., and Aoki, S. (1978). In IUPAC 32nd Int. Symposium Macromol. Prepr., Kyoto, 225.
132. Nikolayev, A. F. and Bondarenko, S. G. (1965). *Vysokomol. Soedin.*, 7:1822.
133. Kataoka, F. and Ando, T. (1981). *Kobunshi Ronbunshu*, 38:821.
134. Nikolayev, A. F. and Bondarenko, S. G. (1970). *Vysokomol. Soedin., Ser. A,* 12:885.
135. Myagchenkov, V. A., Kurenkov, V. F., and Kukushkina, I. A. (1976). *Vysokomol. Soedin., Ser. A*, 19:602.
136. Kurenkov, V. F. and Myagchenkov, V. A. (1979). *Eur. Polym. J.*, 15:849.
137. Kurenkov, V. F., Kukushkina, I. A., and Myagchenkov, V. A. (1978). *Vysokomol. Soedin., Ser. A.*, 20:1271.
138. Kurenkov, V. F., Orlova, G. P., and Myagchenkov, V. A. (1978). *Eur. Polym. J.*, 14:657.
139. Masterova, M. N., Divakar, N. B., Fainstein, A. N., Zubov, V. P., and Kabanov, V. A. (1981). *Vestn. MGU. Khim.*, 22:97.
140. Kurenkov, V. F., Tikhonova, G. P., Kalashnikova, V.I., and Myagchenkov, V.A. (1980). *Vysokomol. Soedin., Ser. A.*, 22:303.
141. Kurenkov, V. F., Vagapova, A. K., and Myagchenkov, V. A. (1982). *Eur. Polym. J.*, 18:763.
142. Czernjawski, T., Dynasisnka, H., and Wojtczak, Z. (1983). *Polimery*, 23:78.

143. Deb, P. C. and Samui, A. B. (1979). *Angew. Makromol. Chem.*, 80:137.
144. Kurakina, L. K., Shmeleva, A. N., and Zabotin, K. P. (1984). *Physico-Chemical Foundations of Polymer Synthesis and Processing*, GGU, Gorky, 86; (1983) 110; (1979) 10.
145. Novospasskaya, I. Yu. and Shmeleva, A. N. (1988). In *Physics and Chemistry of Polymer Synthesis and Properties*, GGU, Gorky, 29.
146. Dzhardimaliyeva, G. I., Pomogailo, A. D., Davtyan, S. P., and Ponomarev, V. I. (1988). *Izv. Akad. Nauk SSSR Khim.*, 1531.
147. Wojtczak, Z. and Czernjawski, T. (1989). *Acta Polym.*, 40:409.
148. Chernjawski, T. and Wojtczak, Z. (1989). *Acta Polym.*, 40:442.
149. Alekseyeva, I. A., Abramova, O. I., and Suvorov, A. L. (1979). *Vysokomol. Soedin., Ser B*, 18:31.
150. Alekseyeva, I. A., Abramova, O. I., and Suvorov, A. L. (1975). *Zh. Obshch. Khim.*, 45:856.
151. Bagdasar'yan, Kh. S. (1966). *Theory of Radical Polymerization*, Nauka, Moscow.
152. Gronowski, A. and Wojtczak, Z. (1983). In Abstr. IUPAC Macro'83, Bucharest, Sect. 1, 423.
153. Levitin, I. Ja. and Vol'pin, M. E. (1984). *J. Mol. Catal.*, 23:315.
154. Selenova, B. S., Muidinov, M. R., and Pomogailo, A. D. (1987). In *Abstracts of Papers 17th All-Union Chugaev Conference on Chemistry of Complex Compounds*, Krasnoyarsk, Part 2, 641.
155. Barkalov, I. M. (1980). *Usp. Khim.*, 49:362.
156. Muidinov, M. R., Dzhardimaliyeva, G. I., Selenova, B. S., Pomogailo, A. D., and Barkalov, I. M. (1988). *Izv. UN SSSR Ser. Khim.*, 2507.
157. Parrish, C. F. and Kochanny, G. L. (1968). *Makromol. Chem.*, 115:119.
158. Bowden, M.J., Kennard, C. H. L., O'Donnell, J. H., Sothman, R. D., Isaacs, N. W., and Smith, G. (1978). *J. Macromol. Sci. Chem.*, 12A:63.
159. Naruchi, K., Janaka, S., Higuma, M., Yamamoto, M., and Yamada, K. (1982). *Polymer*, 23:152.
160. Gronowski, A. and Wojtczak, Z. (1983). *J. Therm. Anal.*, 26:233.
161. Baysal, B. M., Erten, H. N., and Ramelov, U. S. (1971). *J. Polym. Sci. Part A-1*, 9:581.
162. Bowden, M. J. and O'Donnell, J. H. (1968). *Macromolecules*, 1:499.
163. Bowden, M. J. and O'Donnell, J. H. (1968). *J. Phys. Chem.*, 72:1577.
164. Bowden, M. J., O'Donnell, J. H., and Sothman, R. D. (1969). *Makromol. Chem.*, 122:186.
165. Bowden, M. J., O'Donnell, J. H., and Sothman, R. D. (1972). *Macromolecules*, 5:269.
166. Sarma, P. R., Tripathi, K. C., and Gupta, H. M. (1980). *J. Polym. Sci., Polym. Chem. Ed.*, 18:2609.
167. Fydelor, P. J., and Charlesby, A. (1969). *J. Polym. Sci.*, 16C:4493.
168. Dzhardimaliyeva, G. I., Zhorin, V. A., Ivleva, I. N., Pomogailo, A. D., and Enikolopyan, N. S. (1986). *Dokl. Akad. Nauk SSSR*, 287:654.
169. Arutyunyan, Kh. A. and Ushakov, E. N. (1990). In *Abstr. of Papers, All-Union Conference on Kinetics of Liquid-Phase Radical Reactions*, Yaroslavl', 160.
170. Dzhardimaliyeva, G. I., Zhorin, V. A., and Pomogailo, A. D. (In Press). *Izv. Akad. Nauk SSSR Ser. Khim.*
171. Hayashi, T., Oda, J., and Emura, N. (1983). *Sci. Rep. Toyo Soda Manuf. Co.*, Ltd., 27:81.
172. Hamoudi, A. and McNeil, I. C. (1978). *Eur. Polym. J.*, 14:177.
173. Rehak, A., Foldes-Berezsnich, T., Czajlik, I., and Tudos, F. (1982). *Eur. Polym. J.*, 18:435.
174. Zil'berman, E. N., Navolokina, R. A., and Kuvarzina, O. N. (1980). *Vysokomol. Soedin., Ser. A*, 22:2006.
175. Brockman, N. L. and Eisenberg, A. (1985). *J. Polym. Sci., Polym. Phys. Ed.*, 23:1145.
176. McCormick, C. L. and Chen, G. S. (1982). *J. Polym. Sci., Polym. Chem. Ed.*, 20:817.
177. Woroszylo, L. and Uhniat, M. (1981). *Polimery*, 26:52.
178. Bondarenko, S. G., Nikolayev, A. F., Dubkova, N. I., and Kochanenkov, V. P. (1978). *Vysokomol. Soedin., Ser. A*, 20:2433.

179. Wojnarowski, T. J. (1967). *Polimery*, 12:509 (1969). 14:155; (1969). 14:207.
180. Wojtszak, Z. and Suchoka-Galas, K. (1982). *Polimery*, 27:425.
181. Wojnarowski, T. J. and Silin, L. (1973). *Polimery*, 18:128.
182. Lukasik, L., Kuran, W., Laguna, B., and Gask, J. (1983). *J. Polym. Sci., Polym. Chem. Ed.*, 21:3333.
183. Prokop'yev, V. P., Utei, B. I., Khazryatova, L. Kh., and Kuznetsov, E. V. (1977). *Vysokomol. Soedin., Ser. B*, 19:222.
184. Gonyukh, E. A., Serova, V. N., Khazryatova, L. Kh., Smirnov, V. S., and Idiyatullin, D. Sh. (1984). In *Chemistry and Technology of Organoelement Compounds,* Kazan', 58.
185. Gonyukh, E. A., Kuznetsov, E. V., Khazryatova, L. Kh., Prokop'yev, V. P., and Akhmerov, M. A. (1984). *Izv. Vuzov. Khim. Khim. Technol.*, 27: 1070.
186. Abe, Y. and Matsumura, S. (1982). *Jukagaku*, 31:386; (1985) 34:202.
187. Kharas, G. B., Porathnam, S., Jiarui, S., Ozcayir, Y., and Blumstein, A. (1985). *Polym. Bull.*, 13:357.
188. Turner, S. S., Wardle, R., and Thaler, W. A. (1984). *J. Polym. Sci., Polym. Chem. Ed.*, 22:2281.
189. Czaja, K. and Nowakowska, M. (1976). *Polimery*, 21:249; (1978). 23:394.
190. Buchachenko, A. L. and Khudyakov, I. V. (1991). *Usp. Khim.*, 60:1105.
191. Zabotin, K. P. and Malysheva, L. V. (1973). In *Proc. Chemistry and Chemical Technology*, Gorky University, 1 (32):47.
192. Ghanem, N. A., Messiha, N. N., Ikladious, N. E., and Shaaban, A. F. (1981). *J. Appl. Polym. Sci.*, 26:97.
193. Messiha, N. N., Ghanem, N. A., Ikladious, N. E., and Shaaban, A. F. (1980). *Eur. Polym. J.*, 16:1047.
194. Messiha, N. N. (1981). *Polymer*, 22:607.
195. Garg, B. K., Corredor, J., and Subramanian, R. V. (1977). *J. Macromol. Sci. Chem.*, 11A:1567.
196. Yamada, B., Yoneno, H., and Otsu, T. (1970). *J. Polym. Sci. A-1*, 8:2021.
197. Samui, A. B. and Deb, P. C. (1983). *Angew. Makromol. Chem.*, 112:15.
198. Zeldin, M. and Lin, J. J. (1985). *J. Polym. Sci., Polym. Chem. Ed.*, 23:2333.
199. Yankov, V. V. Mamedova, S. G., and Rzaev, Z. M. (1983). In *Monomers and Polymers*, Elm, Baku, 106.
200. Ikladious, N. E., Messiha, N. N., and Shaaban, A. F. (1984). *Eur. Polym. J.*, 20:625.
201. Ikladious, N. E., Messiha, N. N., and Shaaban, A. F. (1984). *J. Appl. Polym. Sci.*, 29:509.
202. Rzaev, Z. M., Gurbanov, K. I., Mamedova, S. G., Guseinov, M. M., and Sharipov, G. S. (1984). *Vysokomol. Soedin., Ser. A*, 26:736.
203. Liepins, R., Timmins, M. L., Morozoff, N., and Surles, J. (1985). In *Metal-Containing Polymer Systems,* Plenum Press, New York, 225.
204. Ghanem, N. A., Messiha, N. N., Ikladious, N. E., and Shaaban, A. F. (1979). *Eur. Polym. J.*, 15:823; (1980) 16:339.
205. Rzaev, Z. M., Medyakova, L. V., and Mamedova, S. G. (1983). *Vysokomol. Soedin., Ser. B*, 25:111.
206. Mamedova, S. G., Rzaev, Z. M., Rasulov, Nk. Sh., Ryzaeva, S. Z., and Agaev, U. Kh. (1986). *Vysokomol. Soedin., Ser. B*, 28:512.
207. Ayrey, G., Humphrey, M. J., and Poller, R. C. (1982). *Eur. Polym. J.*, 18:693.
208. Zubov, V. P., Wulf, V., and Konsulov, V. B. (1990). *Vysokomol. Soedin., Ser. A*, 22:1144.
209. Japanese Patent (1983) 58,185,641; *Chem. Abstr.* (1983) 100:2311223w.
210. Gronowski, A. and Wojtczak, Z. (1985). *Acta Polym.*, 36:59.
211. Wojtczak, Z. and Gronowski, A. (1985). *Makromol. Chem.*, 186:139.
212. Czerniawski, T. and Wojtczak, Z. (1983). In Abstr. IUPAC Macro'83, Bucharest, Sect. 1, 430.

213. Czerniawski, T. and Wojtczak, Z. (1984). *Acta Polym.*, 35:443.
214. Wojtczak, Z., Cherniawski, T., and Rozwadowska, B. (1984). *Acta Polym.*, 34:125.
215. Dzhardimaliyeva, G. I. and Pomogailo, A. D. (1991). *Izv. Akad. Nauk SSSR Khim.*, 352.
216. Barabanov, V. P., Bogoyavlenskaya, L. A. (1973). *Izv. Vuzov. Khim. Khim. Tekhnol.*, 16:118.
217. Higgy, E. S. M., Sayyach, S. M., Rashed, I. H., El-Mamoun, E., and Hussein, A. M. (1986). *Acta Polym.*, 37:606.
218. Baidarovtsev, Yu. P., Savost'yanov, V. S., and Ponomarev, A. N. (In press). *Izv. Akad. Nauk SSSR Khim.*
219. Annenkova, V. Z., Annenkova, V. M., Ugryumova, G. S., Egorshin, V. A., Odintsov, V. V., Andreeva, N. A., and Voronkov, M. G. (1985). *Plastmassy*, 3:1.
220. Tripathi, K. C., Sarma, P. R., and Gupta, H. M. (1978). *J. Polym. Sci., Polym. Chem. Ed.*, 16:629.
221. Heher, H. T., Conn, V. R., and Kroker, E. H. (1950). U.S. Patent 2,502,411; *Chem. Abstr.* (1950) 44:6094e.
222. Church, J. A. (1980). *J. Polym. Sci. A-1*, 5:3188.
223. Morin, B. P., Smolyakov, O. I., Tsarevskaya, I. Yu., Morina, M. V., and Rogovin, Z. A. (1978). *Cell. Chem. Technol.*, 12:263.
224. Dragovic-Bozac, V. (1975). *Tehnica. Hem. Ind.*, 24:649.
225. Penenzhik, M. A., Virnik, A. D., and Rogovin, Z. A. (1969). *Vysokomol. Soedin., Ser. B*, 11:245.
226. Aslan, V., Muteanu, D., and Nica, C. (1976). *Romanian Patent* 65,943.
227. McCormick, C. L. and Park, L. S. (1984). *J. Polym. Sci., Polym. Chem. Ed.*, 22:49.
228. Pavlath, A. and Lee, K. S. (1976). *J. Macromol. Sci. Chem.*, 10A:619.
229. Japanese Patent (1983). 5,719,229; *Chem. Abstr.* (1983) 89:130944g.
230. Kabanov, V. Ya., Voronkov, N. A., Kochkin, D. A., and Spitsyn, V. A. (1971). *Dokl. Akad. Nauk SSSR*, 200:628.
231. Shkolnik, S. and Behar, D. (1982). *J. Appl. Polym. Sci.*, 27:2189.
232. Savost'yanov, V. S., Pomogailo, A. D., Kritskaya, D. A., and Ponomarev, A. N. (1985). *Izv. Akad. Nauk SSSR Khim.*, 45.
233. Savost'yanov, V. S., Pomogailo, A. D., Kritskaya, D. A., and Ponomarev, A. N. (1989). *J. Polym. Sci., Polym. Chem.*, 27:1935.
234. Osada, Y. and Ishida, K. (1976). *Makromol. Chem.*, 177:2209.
235. Savost'yanov, V. S., Zhorin, V. A., Dzhardimaliyeva, G. I., Pomogailo, A. D., Dubovitsky, V. A., Topnikov, V. N., Makova, M. K., and Ponomarev, A. N. (1991). *Dokl. Akad. Nauk SSSR*, 318:278.
236. Gur'yanova, E. N., Goldshtein, I. P. and Romm, I. P. (1973). *The Donor-Acceptor Bond, Khimiya*, Moscow.
237. Tazuke, S. and Okamura, S. (1967). *J. Polym. Sci. Part A-1*, 5:1083.
238. Agnew, N. H. and Brown, M. E. (1971). *J. Polym. Sci. Part A-1*, 9:2561.
239. Tazuke, S., Sato, N., and Okamura, S. (1966). *J. Polym. Sci. Part A-1*, 4:2461.
240. Pomogailo, A. D., Matkovsky, P. E., Konovalov, V. P., Beikhold, G. A., and Leonov, I. D. (1969). *Dokl. Akad. Nauk SSSR*, 184:1364.
241. Laing, M. and Horsfield, E. (1969). *Chem. Commun.*, 902.
242. Agnew, N. H. and Brown, M. E. (1974). *J. Polym. Sci., Polym. Chem.*, 12:1493.
243. Karklin', L. N., Klyuyev, M. V., and Pomogailo, A. D. (1983). *Kinet. Katal.*, 24:408.
244. Skvortsova, G. G., Domnina, E. S., Ivlev, Yu. N., Chipanina, N. N., Skorobogatova, V. I., and Myachin, Yu. A. (1972). *Zh. Obshch. Khim.*, 42:596.
245. Domnina, E. S., Skvortsova, G. G., Glazkova, N. P., Chipanina, N. N., Taryashinova, D. D., and Protasova, L. E. (1976). *Zh. Obshch. Khim.*, 46:168.
246. Domnina, E. S., Voropaev, V. N., Skvortsova, G. G., Sigalov, M. V., Voropaeva, T. K., and Muraveiskaya, G. S. (1983). *Koord. Khim.*, 9:1101.

247. Kalmykov, V. A., Pankov, D. N., and Krychko, E. N. (1979). *Zh. Obshch. Khim.*, 49:647.
248. Domnina, E. S., Ivlev, Yu. N., Shergina, N. I., Chipanina, N. N., Belousova, L. V., Frolov, Yu. L., and Skvortsova, G. G. (1971). *Zh. Obshch. Khim.*, 41:1102.
249. Shchelokov, R. N., Muraveiskaya, G. S., Voropaev, V. N., Skvortsova, G. G., and Domnina, E. S. (1982). *Koord. Khim.*, 8:513.
250. Domnina, E. S., Teterin, Yu. A., Baikalova, L. V., and Skvortsova, G. G. (1986). *Koord. Khim.*, 12:175.
251. Voronov, V. K., Ivlev, Yu. N., Domnina, E. S., and Skvortsova, G. G. (1972). *Khim. Geterotsikl. Soedin.*, 994.
252. Voronov, V. K., Moskovskaya, T. E., Glukhikh, V. I., Rezvukhin, A. N., Domnina, E. S., and Skvortsova, G. G. (1979). *Koord. Khim.*, 5:218.
253. Sokol, V. I., Nikolaev, V. P., Butman, L. A., Poraj-Koshits, M. A., Domnina, E. S., Skvortsova, G. G., and Makhno, L. P. (1979). *Zh. Neorg. Khim.*, 24:727.
254. Kalmykov, V. V., Moiseev, D. A., and Ivanov, V. A. (1985). *Zh. Obshch. Khim.*, 55:1178.
255. Tazuke, S. and Okamura, S. (1969). *J. Polym. Sci. A-1*, 7:851.
256. Voropaev, V. N., Domnina, E. S., Skvortsova, G. G., Teterin, Yu. A., Bayev, A. S., and Voropaeva, T. K. (1984). *Koord. Khim.*, 10:1543.
257. Kalmykov, V. V., Moiseev, D. I., and Ivanov, V. A. (1986). *Koord. Khim.*, 13:730.
258. Pujari, P. and Dash, K. C. (1976). *J. Inorg. Nucl. Chem.*, 38:2183.
259. Broekema, R. J., Durville, P. F. M., Reedijk, J., and Smit, J. A. (1982). *Transition Met. Chem.*, 7:25.
260. Mohapatra, C. K. C. and Dash, K. C. (1977). *J. Inorg. Nucl. Chem.*, 39:1253.
261. Skvortsova, G. G., Es'kova, L. A., Domnina, E. S., Voronov, V. K., Olivson, A. I., Chipanina, N. N., Shulunova, A. M., and Mabarakshina, N. S. (1982). *Koord. Khim.*, 8:1629.
262. Voropaev, V. N., Domnina, E. S., and Skvortsova, G. G. (1986). *Koord. Khim.*, 12:968.
263. Kopylova, L. I., Ivanova, N. D., Voropaev, V. N., Domnina, E. S., Skvortsova, G. G., and Voronkov, M. G. (1985). *Zh. Obshch. Khim.*, 55:1036.
264. Pomogailo, A. D. and Baishiganov, E. B. (1983). In *Complex Organometallic Catalysis for Olefin Polymerization,* Ser. 2, USSR Academy of Sciences, Chernogolovka, 8:66.
265. Goncharov, T. K., Selenova, B. S., and Pomogailo, A. D. (1987). In *Abstr. of Papers 16th All-Union Chugaev Conference on Complex Compound Chemistry*, Krasnoyarsk, Part 1, 571.
266. Gaponik, P. N., Ivashkevich, O. A., and Degtyarik, M. M. (1985). *Izv. Vuzov. Khim. Khim. Tekhnol.*, 11:43.
267. Aliyev, Z. G., Goncharov, T. K., and Grachev, V. P. (1990). *Koord. Khim.*, 16:570.
268. Airis, D. K. (1969). *Carbanions in Organic Synthesis, Khimiya*, Leningrad, 12.
269. Ikegami, T. and Hirai, H. (1970). *J. Polym. Sci. A-1*, 8:195.
270. Matkovsky, P. E., Leonov, I. D., Beikhold, G. A., Pomogailo, A. D., Kissin, Yu. V., and Chirkov, N. M. (1970). *Izv. Akad. Nauk SSSR Khim.*, 1311.
271. Matkovsky, P. E., Leonov, I. D., Kissin, Yu. V., Chirkov, N. M., Pomogailo, A. D., and Beikhold, G. A. (1968). *Izv. Akad. Nauk SSSR Khim.*, 930.
272. Kuran, W., Pasinkiewicz, S., Florjanczcyk, Z. and Lisztyk, E. (1976). *Makromol. Chem.*, 177:2627.
273. Firsov, V. A., Zaitsev, Yu. S., Kucher, R. V., and Kisel, N. G. (1976). In *Physico-Chemical Properties and Synthesis of High-Molecular Compounds*, Naukova Dumka, Kiev, 45.
274. Schrauzer, G. V. (1959). *J. Am. Chem. Soc.*, 81:5310.
275. Kharlamova, E. N., Gur'yanova, E. N., and Slovokhotova, N. A. (1967). *Zh. Obshch. Khim.*, 37:303.
276. Gerard, W., Lappert, M. F., Pyszora, H., and Wallis, J. M. (1960). *J. Chem. Soc.*, 2182.
277. Hirai, H., Ikegami, T., and Makishima, S. (1969). *J. Polym. Sci. A-1*, 7:2059.

278. Hsia Chen, C.S. (1976). *J. Polym. Sci., Polym. Chem. Ed.*, 14:2109.
279. Fowles, W. A. and Gadd, K. F. (1970). *J. Chem. Soc. A.*, 2232.
280. Kern, R. J. (1963). *J. Inorg. Nucl. Chem.*, 25:5.
281. Pomogailo, A. D. and Beikhold, G. A. (1969). *Dokl. Akad. Nauk SSSR*, 188:148.
282. Woo, J. C., Kim, S. H., and Chin, C. S. (1982). *J. Polym. Sci., Polym. Chem. Ed.*, 20:1947.
283. Nametkin, N. S., Sorokin, G. V., Kuz'yants, G. M., Pozdnyakova, M. A., Sulima, V. V., and Sytov, G. A. (1972). *Izv. Akad. Nauk SSSR Khim.*, 1931.
284. Nametkin, N. S., Perchenko, V. N., Sharafutdinov, M. S., and Volkova, V. V. (1974). *Dokl. Akad. Nauk SSSR*, 214:1084.
285. Nametkin, N. S., Sharafutdinov, M. S., Perchenko, V. N., Sideridu, A. Ya., and Sorokin, G. V. (1969). *Dokl. Akad. Nauk SSSR*, 189:1264.
286. Nametkin, N. S., Sharafutdinov, M. S., Perchenko, V. N., Sorokin, G. V., and Galiakbarov, M. F. (1970). *Dokl. Akad. Nauk SSSR*, 192:1049.
287. Sytov, G. A., Krapivin, A. M., Perchenko, V. N., Nametkin, N. S., and Sharafutdinov, M. S. (1977). *Dokl. Akad. Nauk SSSR*, 237:114.
288. Perchenko, V. N., Sytov, G. A., Ledina, L. E., Ishmukhametova, N. K., Krapivin, A. M., and Nametkin, N. S. (1981). *Dokl. Akad. Nauk SSSR*, 261:114.
289. Perchenko, V. N., Sytov, G. A., Kamneva, G. L., and Nametkin, N. S. (1983). *Vysokomol. Soedin., Ser. A*, 25:1723.
290. Nametkin, N. S., Perchenko, V. N., Sytov, G. A., and Mirskova, I. S. (1976). *Dokl. Akad. Nauk SSSR*, 226:98.
291. Ukraintsev, V. B., Yakovlev, S. V., and Kukushkin, Yu. N. (1984). *Zh. Obshch. Khim.*, 54:1592.
292. Pomogailo, A. D., Baishiganov, S., and Khvostik, G. M. (1980). *Kinet. Katal.*, 21:1535.
293. Imoto, M., Otsu, T., and Harada, Y. (1963). *Makromol. Chem.*, 65:180.
294. Ikegami, T. and Hirai, H. (1970). *J. Polym. Sci. A-1*, 8:463.
295. Sergeev, G. B., Komarov, V. S., and Fedorova, A. T. (1982). *Vestn. MGU Khim.*, 23:114.
296. Farona, M. F., Agers, W. T., Ramsey, B. G., and Grasseli, J. G. (1969). *Inorg. Chim. Acta*, 3:503.
297. Barvinok, M. S. and Mashkov, L. V. (1971). *Zh. Neorg. Khim.*, 19:571.
298. Farona, M. F., Grasseli, J. G., Grossman, H., and Ritchey, W. (1969). *Inorg. Chim. Acta*, 3:495.
299. Bugaeva, L. N. and Kryukov, A. I. (1980). *Zh. Neorg. Khim.*, 25:1018.
300. Reedijk, J. (1970). *Inorg. Chim. Acta*, 5:687.
301. Savost'yanov, V. S., Ponomarev, V. I., Selenova, B. S., Pomogailo, A. D., Starikov, A. G., and Atovmyan, L. O. (1990). *Izv. Akad. Nauk SSSR Khim.*, 762.
302. Barvinok, M. S. and Mashkov, L. V. (1978). *Zh. Neorg. Khim.*, 23:1843.
303. Uflyand, I. E., Starikov, A. G., Savost'yanov, V. S., and Pomogailo, A. D. (1990). *Izv. Akad. Nauk SSSR Khim.*, 1319.
304. Uflyand, I. E., Starikov, A. G., Savost'yanov, V. S., and Pomogailo, A. D. (1991). *Transition Met. Chem.*, 16:126.
305. Zurakowska-Orszagh, J., Mirowski, K., and Kaim, A. (1984). *J. Polym. Sci., Polym. Lett. Ed.*, 22:649.
306. Zurakowska-Orszagh, J., Mirowski, K., and Kaim, A. (1981). *J. Polym. Sci., Polym. Chem. Ed.*, 19:363.
307. Kilic, S. and Baysal, B. M. (1984). *J. Polym. Sci., Polym. Chem. Ed.*, 22:453.
308. Zurakowska-Orszagh, J., Mirowski, K., and Chajewski, A. (1982). *J. Polym. Sci., Polym. Chem. Ed.*, 20:1775.
309. Kalayeva, M. I. and Akhmedova, Kh. A. (1982). In *Physico-Chemical Methods of Analysis and Production Control*, Makhachkala, 67.
310. Nikonov, V. A., Leplyalin, G. V., Murinov, Yu. I., and Derzhinsky, A. R. (1984). *Izv. Akad. Nauk SSSR Khim.*, 2647.

311. Nikonov, V. A., Kurochkin, A. V., and Leplyanin, G. V. (1989). *Izv. Akad. Nauk SSSR Khim.*, 1504.
312. Ragg, P. L. (1974). U.S.S.R. Patent, 445,203; *Bull. Izobr.* (1971). 36:146.
313. Fyfe, C. A., Clark, H. C., Davies, J. A., Hayes, P. J., and Wasylishen, R. E. (1983). *J. Am. Chem. Soc.*, 105:6577.
314. Keiter, R. L., Sun, Y. Y., Brodack, J. W., and Cary, L.W. (1979). *J. Am. Chem. Soc.*, 101:2638.
315. Treuchel, P. M. and Wong, W. K. (1979). *Inorg. Chim. Acta*, 33:171.
316. Bamford, C. H., Jenkins, A. D., and Jonston, R. (1957). *Proc. R. Soc. London, Sect. A*, 241:364.
317. Kargin, V. A., Kabanov, V. A., and Zubov, V. P. (1960). *Vysokomol. Soedin.*, 2:765.
318. Imoto, M., Otsu, T., and Shimizu, S. (1963). *Makromol. Chem.*, 65:174.
319. Ragg, P. L. (1970). German Patent, 1,937,225; *Chem. Abstr.* (1970). 72:79628d.
320. Danilovtseva, E. N., Skushnikova, A. I., Domnina, E. S., and Afonin, A.V. (1989). *Vysokomol. Soedin., Ser. B*, 31:777.
321. Skushnikova, A. I., Danilovtseva, E. N., Domnina, E. S., and Vakul'skaya, T. I. (1990). *Vysokomol. Soedin., Ser. A*, 32:2331.
322. Skvortsova, G. G., Domnina, E. S., Ivlev, Yu. N., and Chipanina, N. N. (1976). *Zh. Obshch. Khim.*, 46:1593.
323. Skushnikova, A. I., Domnina, E. S., and Skvortsova, G. G. (1982). *Vysokomol. Soedin., Ser. B*, 24:11.
324. Ivlev, Yu. N., Domnina, E. S., Skvortsova, G. G., and Shostakovsky, M. F. (1969). U.S.S.R. Patent, 271,014; *Bull. Izobr.* (1970). 17:78.
325. Skvortsova, G. G., Domnina, E. S., Voropaev, V. N., Chipanina, N. N., Sokol., V. I., and Teterin, Yu. A. (1985). In Abstr. Papers 5th All-Union Conference on the Chemistry of Aqueous Solutions of Inorganic and Complex Compounds, Rostov-on-Don, 243.
326. Maekawa, T., Mah, S., and Okamura, S. (1978). *J. Macromol. Sci.-Chem.*, 12A:1.
327. El-Sonbati, A. Z., Diab, M. A., Hilali, A. S., and Moussa, M. N. (1989). *Acta Polym.*, 40:236.
328. Diab, M. A., El-Sonbati, A. Z., Hilali, A. S., Killa, H. M., Ghoneim, M. M. (1990). *Eur. Polym. J.*, 26:1.
329. Savost'yanov, V. S., Vasilets, V. N., Ermakov, O. V., Sokolov, E. A., Pomogailo, A. D., Kritskaya, D. A., and Ponomarev, A. N. (1992). *Izv. Akad. Nauk SSSR Khim.*, 2073.
330. Savost'yanov, S. V., Pilyugin, V.V., Kritskaya, D.A., and Ponomarev, A.N. (1993): *Izv. Akad. Nauk SSSR Khim.*, 1070.
331. Kabanov, D. A., Zubov, V. P., and Semchikov, Yu. D. (1987). *Complex-Radical Polymerization*, Khimiya, Moscow.
332. Ganzyuk, L. I., Stepanova, L. S., and Chernysheva, L. S. (1976). *Vysokomol. Soedin, Ser. B*, 18:101.
333. Okuzawa, S., Hirai, H., and Makishima, S. (1969). *J. Polym. Sci. A-1*, 7:1039.
334. Zurakowska-Orszagh, J. and Bartnic, J. (1986). *Polimery*, 31:153.
335. Zurakowska-Orszagh, J., Mirowski, K., and Wilczura, H. (1981). *J. Therm. Anal.*, 21:27.
336. Savost'yanov, V.S., Pomogailo, A.D., Selenova, B. S., Kritskaya, D. A., and Ponomarev, A. N. (1990). *Izv. Akad. Nauk SSSR Khim.*, 768.
337. Savost'yanov, V. S., Belov, G. P., Kritskaya, D. A., Pomogailo, A. D., and Ponomarev, A. N. (1990). *Izv. Akad. Nauk SSSR Khim.*, 1068.
338. Savost'yanov, V. S., Pomogailo, A. D., Kritskaya, D. A., and Ponomarev, A. N. (In press). *J. Polym. Sci., Polym. Chem.*
339. Kabanov, V. A. (1967). *Usp. Khim.*, 36:217.
340. Enikolopyan, N. S., Bochkin, A. M., Pomogailo, A. D., Kuzayev, A. I., Ivanchenko, P. A., Ivleva, I. N., and D'yachkovsky, F. S. (1982). *Dokl. Akad. Nauk SSSR*, 265:95.
341. Li, T. and Zhou, C. (1989). *Acta Polym. Sin.*, 3:316.

342. Kato, M., Nishide, H., Tsuchida, E., and Sasaki, T. (1981). *J. Polym. Sci., Polym. Chem. Ed.*, 19:1803.
343. U.S. Patent (1984) 4,474,688.
344. Komiyama, M. and Hirai, H. (1976). *J. Polym. Sci., Polym. Chem. Ed.*, 14:627.
345. Cabasso, I., Smid, J., and Sahni, S.K. (1989). *J. Appl. Polym. Sci.*, 38:1653.
346. Pomogailo, A. D., Leonov, I. D., and Afanas'yeva, I. I. (1970). *Vestn. Akad. Nauk Kazakh. SSR*, 10:29.
347. Pomogailo, A. D., Baishiganov, E., and D'yachkovsky, F. S. (1981). *Vysokomol. Soedin., Ser. A*, 23:220.
348. Dubinina, L. M., Amerik, V. V., and Kleiner, V. I. (1983). In *Homo- amd Copolymerization of Olefins Using Complex Catalysts*, Nauka, Moscow, 33.
349. Pomogailo, A. D., Matkovsky, P. E., Zavorokhin, N. D., Sokol'sky, D. V., and Beikhol'd, G. A. (1967). *Dokl. Akad. Nauk SSSR*, 176:1347.
350. Furukawa, J. and Iseda, Y. (1969). *J. Polym. Sci.*, 7B:47.
351. Ivantchev, S. S., Spevak, L. L., Ivantcheva, N. I., and Khaikin, S. Ya. (1984). *Vysokomol. Soedin., Ser. B*, 26:517.
352. Czaja, K. and Kurzawa, M. (1985). *Polimery*, 30:224.
353. George, A., Radhakrishnan, G., and Joseph, K. T. (1984). *J. Macromol. Sci. Chem.*, 21A:179.
354. Kuramochi, T., Kawamura, K., Sugita, S., and Ibonai, M. (1982). *J. Chem. Soc. Jpn. Chem. Ind. Chem.*, 670.
355. Kreitsberg, G. N., Tsailingold, V. L., Smirnova, N. I., Petrushanskaya, N. V., Fel'dblyum, V. Sh., Kopylov, E. P., Stepanov, G. A., Ustavshchikov, O. E., Kabanov, V. A., Smetanyuk, V. I., Prudnikov, A. I., and Pluzhnov, S. K. (1981). U.S.S.R. Patents 827251–827153; *Chem. Abstr.* (1981). 95:149923x, 149924y, 61454j.
356. Plate, N. A., Valuev, L. I., Masterova, M. N., and Istomin, A. P. (1978). U.S.S.R. Patent 608,813; *Chem. Abstr.* (1978) 89:758536.
357. Kurbanov, Sh. A., Khakimdzhanov, B. Sh., Musaev, U. N., and Tillaev, R. S. (1975). In Col. Research Works, Tashkent University, No. 498, 26.
358. Savost'yanov, V. S., Pomogailo, A. D., Kritskaya, D. A., and Ponomarev, A. N. (1986). *Izv. Akad. Nauk SSSR Khim.*, 42.
359. Savost'yanov, V. S., Pomogailo, A. D., Kritskaya, D. A., and Ponomarev, A.N. (1986). *Izv. Akad. Nauk SSSR Khim.*, 353.
360. Uflyand, I. E. and Pomogailo, A. D. (1991). *Usp. Khim.*, 60:1532.
361. Kudaibergenov, S. E., Saikhutdinov, E. M., Zhaimina, G. M., and Bekturov, E. A. (1985). *Koord. Khim.*, 11:1329.
362. Potts, K. T., Usifer, D. A., Guadelupe, A., and Abruna, H. D. (1987). *J. Am. Chem. Soc.*, 109:3951.
363. Ghosh, P. K. and Spiro, T. G. (1980). *J. Am. Chem. Soc.*, 102:5543.
364. Ennis, P. A., Kelly, J. M., and O'Connell, C. M. (1986). *J. Chem. Soc. Dalton Trans.*, 2485.
365. Meyer, T. J., Sullivan, B. P., and Caspar, J. V. (1987). *Inorg. Chem.*, 26:4145.
366. Sumi, K., Kurue, M., and Nozakura, S.-I. (1985). *J. Polym. Sci., Polym. Chem. Ed.*, 23:359.
367. Baikalova, L. V., Domnina, I. S., Skvortsova, G. G., Protasova, L. E., Chipanina, N. N., and Voronov, V. K. (1982). *Koord. Khim.*, 8:1038.
368. Domnina, E. S., Teterin, Yu. A., Baikalova, L. V., and Skvortsova, G. G. (1986). *Koord. Khim.*, 12:175.
369. Sato, T., Morita, N., Kamiya, I., Tanaka, H., and Ota, T. (1990). *Makromol. Chem. Rapid Commun.*, 11:553.
370. Diab, M. A., El-Sonbati, A. Z., El-Bindary, A. A., Aggour, A. A. (1990). *J. Therm. Anal*, 36:957.
371. Clear, J. M., Kelly, J. H., and O'Connell, C. M. (1981). *J. Chem. Res. Part M*, 30/34:3039.

372. Calvert, I. M., Schemhl, R. H., Sullivan, B. P., Facci, J. S., Meyer, T. J., and Murray, R. W. (1983). *Inorg. Chem.*, 22:2151.
373. Abruna, H. D., Denisevich, P., Umana, M., Meyer, J., and Murray, R. W. (1981). *J. Am. Chem. Soc.*, 103:1.
374. Zarnegar, P. P. and Whitten, D. G. (1971). *J. Am. Chem. Soc.*, 93:3776.
375. Powers, M. J., Gallahan, R. W., Salmon, D. J., and Meyer, T. J. (1976). *Inorg. Chem.*, 15:894.
376. Uflyand, I. E., Kokoreva, I. V., Starikov, A. G., Sheinker, V. N., and Pomogailo, A. D. (1989). *Izv. Akad. Nauk SSSR Khim.*, 2468.
377. Zagnii, V. V., Syromyatnikov, V. G., and Sinyavsky, N. G. (1987). *Vestn. Kiev Univ. Khim.*, 28:37.
378. Teyssie, P. and Smets, G. (1958). *Makromol. Chem.*, 26:245.
379. Yeh, H. C., Eichinger, B. E., and Andersen, N. H. (1982). *J. Polym. Sci., Polym. Chem. Ed.*, 20:2575.
380. Voloshanovskii, I. S., Movchan, T. I., and Pakatina, L.Yu. (1990). In Nontraditional Methods of Polymer Synthesis, Int. School Semin., Alma-Ata, 164.
381. Okamoto, Y., Wang, S. S., and Zhu, K. J. (1985). In *Metal-Containing Polymer Systems*, Plenum Press, New York, 425.
382. Bindlish, J. M., Bhatia, S. G., and Gantam, P. (1978). *Ind. J. Chem.*, 16A:279.
383. Tsuchida, E., Tomono, T., and Honda, K. (1974), *J. Polym. Sci., Polym. Chem. Ed.*, 12:1243.
384. Tsuchida, E., Tomono, T., and Honda, K. (1973). *J. Chem. Soc. Jpn., Chem. Ind. Chem.*, 1184.
385. Uflyand, I. E., Il'chenko, I. A., Starikov, A. G., Sheinker, V. P., and Pomogailo, A. D. (1989). *Izv. Akad. Nauk SSSR Khim.*, 2474.
386. Ueba, Y., Zhu, K. J., Banks, E., and Okamoto, Y. (1982). *J. Polym. Sci., Polym. Chem. Ed.*, 20:1271.
387. Potapov, G. P. and Aliyeva, M. I. (1983). *Izv. Vuz., Khim. Khim. Tekhnol.*, 26:1122.
388. Kamogawa, H. (1972). *J. Polym. Sci., Polym. Lett. Ed.*, 20:711.
389. Solov'eva, A. B., Samokhvalova, A. I., Lebedeva, T. S., Pshezhetsky, V. S., Kormilova, L. V., and Enikolopyan, N. S. (1986). *Dokl. Akad. Nauk SSSR*, 290:1383.
390. Itoh, H., Kondo, S., Masuda, E., Hanabusa, K., Shirai, H., and Nojo, N. (1986). *Makromol. Chem. Rapid. Commun.*, 7:585.
391. Kamachi, M., Cheng, X.S., and Nozakura, S. (1987). In Tokyo Seminar on Macromolecule-Metal Complexes, Prepr., Tokyo, 41.
392. Ellis, C. D., Margerum, L. D., Murray, R. W., and Meyer, T. J. (1983). *Inorg. Chem.*, 22:1283.
393. Elliott, C. M., Redepenning, J. G., and Balk, E. M. (1986). *J. Electroanal. Chem.*, 203:1013.
394. Elliott, C. M., Redepenning, J. G., and Balk, E. M. (1985). *J. Am. Chem. Soc.*, 107:8302.
395. Elliott, C. M. and Redepenning, J. G. (1986). *J. Electroanal. Chem.*, 197:219.
396. Fujii, Y., Kikuchi, K., Matsutani, K., and Ota, K. (1984). *Chem. Lett.*, 1487.
397. Nishide, H., Shinohara, K., and Tsuchida, E. (1981). *J. Polym. Sci., Polym. Chem. Ed.*, 11:1109.
398. Kokufuta, E., Vatanabe, H., and Nakamura, I. (1981). *Polym. Bull.*, 4:603.
399. Furhhop, J. H., Besecke, S., Vogt, W., Ernst, J., and Subramanian, J. (1977). *Makromol. Chem.*, 178:1621.
400. Nishide, H., Kato, M., and Tsuchida, E. (1981). *Eur. Polym. J.*, 17:579.
401. Kamogawa, H. (1974). *J. Polym. Sci., Polym. Chem. Ed.*, 12:2317.
402. Uflyand, I. E., Pomogailo, A. D., and Sheinker, V. N. (1987). In Proc. 25th Int. Conf. Coordination Chem., Tanjing, China, 497.
403. Uflyand, I. E., Kokoreva, I. V., and Starikov, A. G. (1989). *React. Polym.*, 11:211.
404. Uflyand, I. E., Il'chenko, I. A., and Sheinker, V. N. (1991). *React. Polym.*, 14:41.
405. El-Bindari, A. A., Ed. *Transition Metal Clusters*, Wiley Interscience, New York.

406. Deeming, A. J. and Hasso, S. (1976). *J. Organomet. Chem.*, 114:313.
407. Korniets, E. D., Kedrova, L. K., and Pavlenko, N. I. (1986). *Izv. Akad. Nauk SSSR Khim.*, 2310.
408. Dzhardimaliyeva, G. I., Pomogailo, A. D., and Tunik, S. P. (In press). *Izv. Akad. Nauk SSSR Khim.*
409. Kaganovich, V. S., Rybinskaya, M. I. (1987). *Izv. Akad. Nauk SSSR Khim.*, 441.
410. Sheats, J. E., Carraher, C. E., Jr. and Pittman, C. U., Jr., Eds., (1985). *Metal-Containing Polymer Systems*, Plenum Press, New York, 291.
411. Gressier, J. C., Levesque, G., and Patin, H. (1982). *Polym. Bull.*, 8:55.
412. Bravaya, N. M. and Pomogailo, A. D. (In press). *Izv. Akad. Nauk SSSR Khim.*
413. Bovey, F. A. (1972). *High-Resolution NMP of Macromolecules,* Academic Press, New York.
414. Pokrovsky, E. I. (1964). *Vysokomol. Soedin.*, 6:642.
415. Monjol, P. and Champetier, G. (1972). *Bull. Soc. Chim. Fr.*, 1302.
416. Chapiro, A., Goldfeld-Freichlich, D., and Perichon, J. (1975). *Eur. Polym. J.*, 11:515.
417. Chirkov, N. M. (1970). *Kinet. Katal.*, 11:321.
418. Bovey, F. A. (1963). *J. Polym. Sci.*, 1A:843.
419. Kakivaya, S. R. and Salamone, J. C. (1974). *J. Polym. Sci., Polym. Lett. Ed.*, 12:651.
420. Ledsham, R. C. and Day, P. (1981). *J. Chem. Soc. Chem. Commun.*, 921.
421. Selenova, B. S., Raspopov, L. N., Savost'yanov, V. S., and Pomogailo, A. D. (1991). In, *Complex Organometallic Compounds for Olefin Polymerization*, OIKhF AN SSSR, Chernogolovka, 11, 115.
422. Botsman, A. V. and Marchenko, A. V. (1986). *Ukr. Khim. Zh.*, 52:952.
423. Volkov, S. V., Botsman, A. V., Lutoshkin, V. I., and Pomogailo, A. D. (1988). *Ukr. Khim. Zh.*, 54:899.
424. Botsman, A. V., Volkov, S. V., Lutoshkin, V. I., and Pomogailo, A. D. (1989). *Ukr. Khim. Zh.*, 55:341.
425. Klein, L. and Scholz, W. (1979). *Makromol. Chem.*, 180:1477.
426. Skazka, V. S. and Tarasova, G. V. (1978). *Vysokomol. Soedin., Ser. B*, 20:97.
427. Feiffer, D. G. and Lundberig, R. D. (1984). *J. Polym. Sci., Polym. Chem. Ed.*, 22:1757.
428. Vorreux, G., Morcellet, M., and Loucheux, C. (1982). *Makromol. Chem.*, 183:711.
429. Kulicke, W. M. and Kniewske, R. (1981). *Makromol. Chem.*, 182:2277.
430. Suita, T. (1980). *Jpn. J. Appl. Phys.*, 19:2501.
431. Ise, N., Okubo, T., Hiragi, Y., Kawai, H., Hashimoto, T., Fujimura, M., Nakajima, A., and Hayashi, H. (1979). *J. Am. Chem. Soc.*, 101:5836.
432. Eisenberg, A. (1970). *Macromolecules*, 3:147.
433. Eisenberg, A. (1976). *Pure Appl. Chem.*, 46:171.
434. Suchocka-Galas, K. (1983). *Polimery*, 28:260.
435. Neppel, A., Butler, I.S., and Eisenberg, A. (1979). *J. Polym. Sci., Polym. Phys. Ed.*, 17:2145.
436. Eisenberg, A., Ridahl, M., and Brockman, N. (1980). In IUPAC Macro'80 Int. Symp. Macromolecules, Prepr., Pisa, Vol. 3, 167.
437. Bagrodia, S., Pisipati, R., Wilkes, G. L., Robson, F., Kennedy, S. P., and Kennedy, J. P. (1984). *J. Appl. Polym. Sci.*, 29:3065.
438. Piner, M., Duplessix, R., Gautieir, S., and Eisenberg, A. (1980). In *Ion Polymer Symp. 176th Miami Beach, 1976*, U.S. Government Printing Office, Washington, D.C., 283.
439. Takahashi, T. and Hara, K. (1983). *Kobunshi Ronbunshu*, 40:713.
440. Utracki, L. A. (1982). *Polym. Eng. Sci.*, 22:81, 85.
441. Gressier, J. C., Levesque, G., and Patin, A. (1982). *Polym. Bull.*, 8:55.
442. Ouchi, T., Inaba, M., Tadano, K., and Imoto, M. (1982). *J. Polym. Sci., Polym. Chem. Ed.*, 20:2089.
443. German Patent (1981) 3,137,840.
444. Sosnowski, S., Stomkowski, S., and Penczek, S. (1987). *Makromol. Chem.*, 186:1347.

445. Tazuke, S. and Okamura, S. (1966). *J. Polym. Sci., A-1*, 4:141.
446. Kalmykov, V.V., Ivanov, V.A., and Kharitonov, G.V. (1979). *Zh. Prikl. Khim.* (Leningrad), 51:889.
447. Pomogailo, A.D. (1991). *Catalysis with Immobilized Metallocomplexes*, Nauka, Moscow.
448. Dzhardimaliyeva, G.I., Ivanchenko, P.A., and Pomogailo, A.D. (1984). In Abstr. Papers 4th Int. Symp. Homogeneous Catalysis, Leningrad, 1117.
449. Uflyand, I. E., Pomogailo, A. D., Kokoreva, I. V., Dzhardimaliyeva, G. I., Khrisostomov, F. A., and Sheinker, V. A. (1988). U.S.S.R. Patent, 1,512,650; *Byull. Izobr.* (1989). 37:34.
450. Savost'yanov, V. S., Pomogailo, A. D., and Ponamarev, A. N. (1989). *Kinet. Katal.*, 30:1414.
451. Dzhardimaliyeva, G. I., Selenova, B. S., and Pomogailo, A. D. (1986). In Abstr. Papers 5th Int. Symp. Relation between Homogeneous and Heterogeneous Catalysis, Novosibirsk, 78.
452. Macomber, D. W., Hart, W. P., Rausch, M. D., Priester, R. D., and Rittman, C. U., Jr. (1982). *J. Am. Chem. Soc.*, 104:884.
453. Pittman, C. U., Jr. (1980). *Polym. News*, 6:169.
454. Pomogailo, A. D. and Klyuev, A. V. (1985). *Izv. Akad. Nauk SSSR Khim.*, 1716.
455. Uflyand, I. E., Il'chenko, I. A., and Sheinker, V. N. (1991). *React. Polym.*, 14:160.
456. Nikitin, A. V., Pomogailo, A. D., and Rubailo, V. L. (1987). *Izv. Akad. Nauk SSSR Khim.*, 36.
457. Nikitin, A. V., Pomogailo, A. D., Maslov, S. A., and Rubailo, V. L. (1987). *Neftekhimiya*, 27:234.
458. Carraher, C. E., Jr., Ademie, Y. C., and Giron, D. J. (1985). In *Metal-Containing Polymer Systems*, Plenum Press, New York, 197.
459. Parmon, V. N. (1984). Development of Physico-Chemical Foundations for the Conversion of Solar Energy by Water Decomposition in Molecular Photocatalytic Systems, Diss., Novosibirsk.
460. Ennis, P. M., Kelly, J. M., and O'Connell, C. M. (1986). *J. Chem. Soc. Dalton Trans.*, 2485.
461. Kaneko, M., Yamada, A., Oyama, N., and Yamaguchi, S. (1982). *Makromol. Chem. Rapid. Commun.*, 3:769.
462. Tanaka, T., Kaneko, M., and Yamada, A. (1980). In IUPAC Macro'80 Int. Symp. Macromolecules, Prepr., Pisa, Vol. 4, 103.
463. Kamogawa, H., Inoue, H., and Nansawa, M. (1980). *J. Polym. Sci., Polym. Chem. Ed.*, 18:2209.
464. U.S.S.R. Patent (1981) 1,021,456.
465. Kopylova, L. I., Ivanova, N. D., Voropaev, V. N., Domnina, E. S., Skvortsova, G. G., and Voronkov, M. G. (1985). *Zh. Obshch. Khim.*, 55:1036.
466. *Marine Growth and Biocorrosion Processes* (1981). Nauka, Moscow.
467. Mishchenko, V. F., Zubov, V. P., and Eremenko, Yu. G. (1985). *Vysokomol. Soedin., Ser. B*, 27:341.
468. Subramanian, R. V. and Somasekharan, K. N. (1981). *J. Macromol. Sci. Chem.*, 16A:73.
469. Zhiltsov, S. F. and Kashayeva, V. N. (1984). *Physico-chemical Foundations of Polymer Synthesis and Processing*, Gorky, 66.
470. Japanese Patent (1983) 121,212.
471. Annenkova, V. Z., Platonova, A. T., Kononchuk, G. M., Dinova, N. G., Kazimirovskaya, V. B., Annenkova, V. M., Ugryumova, G. S., and Voronkov, M. G. (1982). *Khim.-Farmatsevt. Zh.*, 3:66.
472. German Patent (1983) 3,305,121.
473. U.S. Patent (1977) 4,321,273.
474. Krzyszowska, K. (1975). In 4th Int. Conf. Thermal Analysis, Budapest. Vol. 2, 159.

475. McNeil, I. C. and Zulfigar, M. (1978). *J. Polym. Sci., Polym. Chem. Ed.*, 16:3201.
476. Serova, V. N., Kuznetsov, E. V., and Khazryatova, L. Kh. (1983). Dep. Manusc. ONIITEkhim, No. 1102khp-D83.
477. Gronowski, A. and Wojtczak, Z. (1989). *Acta Polym.*, 40:265.
478. Rozwadowska, B., Growovski, A., and Wojtczak, Z. (1990). *J. Therm. Anal.*, 36:939.
479. Dzhardimaliyeva, G. I., Selenova, B. S., Ponomareva, T. I., Skachkova, V. K., and Pomogailo, A. D. (1986). In Complex Organometal Catalysts for Olefin Polymerization. OIKhF AN SSSR, Chernogolovka, 10, 55.
480. Gronowski, A. and Wojtczak, Z. (1985). *J. Therm. Anal.*, 30:345.
481. Szalinska, H., Pietrzak, M., and Janowska, G. (1986). *Radiat. Phys. Chem.*, 27:101.
482. Japanese Patent (1977) 53,115,766.
483. Japanese Patent (1976) 5,312,948.
484. U.S. Patent (1969) 3,705,137.
485. U.S. Patent (1977) 4,146,698.
486. U.S. Patent (1979) 4,178,423.
487. Japanese Patent (1972) 5,023,867.
488. Serova, V. N., Kuznetsov, E. V., and Khazryatova, L. Kh. (1985). In *Physico-Chemical Foundation of Polymer Synthesis and Processing*, Gorky, 37.
489. Antonova, L. F., Ionova, I. L., Leplyanin, G. V., and Rafikov, S. R. (1975). Dep. Manuscr. VINITI 3026-75; *Ref. Zh. Khim.* (1976) 1S216.
490. Romanko, O. I., Zakharova, N. N., Andreeva, I. N., Volkov, L. A., Kalashnik, A. T., and Zverev, M. P. (1982). *Khim. Volokna*, 3:17.
491. Liepins, R., Timmons, M. L., Morozoff, N., and Surles, J. (1985). In *Metal-Containing Polymer Systems*, Plenum Press, New York, 225.
492. Utei, B. I., Zuev, B. M., Akhmatgaliyeva, L. Kh., and Kuznetsov, E. V. (1975). In *Proc. Kazansky Chemical-Technol. Inst.*, 56:130.
493. U.S. Patent (1960) 2,940,957.
494. Japanese Patent (1981) 57,139,034.
495. German Patent (1979) 2,943,566.
496. German Patent (1981) 3,114,266.
497. Korshak, V. V. (1970). *Chemical Structure and Temperature Characteristics of Polymers*, Nauka, Moscow.
498. Webb, D. J. and Hatzakis, M. (1979). *J. Vac. Sci. Techn.*, 16:1008.
499. U.S. Patent (1979) 4,156,745.
500. Mamedova, S. G., Rzaev, Z. M., Rasulov, N. Sh., Rizaeva, S. Z., and Agaev, U. Kh. (1986). *Vysokomol. Soedin., Ser. B*, 28:512.
501. U.S. Patent (1981) 44,290,194.
502. Japanese Patent (1982) 56,185,641.
503. Japanese Patent (1978) 5,573,717.
504. Japanese Patent (1978) 5,495,695.
505. Margerum, L. D., Meyer, T. J., and Murray, R. W. (1983). *J. Electroanal. Chem.*, 154:279.
506. Serova, V. N., Gonyukh, S. A., Khazryatova, L. Kh. (1984). In *Chemistry and Technology of Organoelement Compounds and Polymers*, Kazan', 56.
507. Pomogailo, A. D., Savost'yanov, V. S., Dubovitsky, A. V., and Ponomarev, A. N. (In press). *Izv. Akad. Nauk RAN.*
508. U.S.S.R. Patent (1975) 570,618.
509. Japanese Patent (1979) 5,676,419.
510. Japanese Patent (1980) 56,161,408.
511. Japanese Patent (1982) 5,918,712.
512. Japanese Patent (1979) 5,540,749.
513. Japanese Patent (1980) 56,161,414.
514. Japanese Patent (1979) 5,684,632.

515. Japanese Patent (1980) 578,210.
516. U.S. Patent (1974) 3,912,904.
517. U.S. Patent (1978) 4,167,464.
518. Japanese Patent (1979) 5,670,011.
519. Japanese Patent (1972) 5,023,858.
520. U.S. Patent (1982) 4,459,396.
521. German Patent (1978) 3,124,008.
522. U.S.S.R. Patent (1979) 682,531.
523. U.S. Patent (1979) 4,288,582.
524. Sadykov, R. M., Samoilova, G. G., Min'kov, V. A., Gubanova, T. S., and Zkordonets, S. P. (1984). *Lakokras. Mat. Ikh. Primen.*, 5:11.
525. French Patent (1971) 2,133,134.
526. U.S. Patent (1969) 3,479,381.
527. U.S. Patent (1979) 4,309,329.
528. Japanese Patent (1979) 5,688,407.
529. Japanese Patent (1972) 5,430,021.
530. Japanese Patent (1982) 50,108,074.
531. Polish Patent (1978) 115,019.
532. Polish Patent (1976) 105,454.
533. German Patent (1982) 3,248,019.
534. Ustinov, V. V., Mednikov, M. M., and Imin, I. A. (1979). In Abstr. Papers All-Union Scientific and Technological Conference on Methods of Elastomer Modification for Rubber Product Quality Improvement, Manufacturing Efficiency, Yaroslavl', 115.
535. Japanese Patent (1975) 5,412,448.
536. Japanese Patent (1976) 52,132,039.
537. Kochkin, D. M., Kotrelev, V. N. Kalinina, S. I., Kuznetsova, G. I., Laine, L. V., Chernova, L. V., Borisova, A. I., and Borisenko, V. V. (1959). *Vysokol. Soedin.*, 1:1507.
538. Uflyand, I. E. and Kuzharov, A. S. (1990). In Bezyznosnost. Interinstitution Scientific Collected Papers, Rostov-on-Don, 132.
539. Uflyand, I. E., Kuzharov, A. S., and Pomogailo, A. D. (1987). Proc. 25th Int. Conf. Coordination Chemistry, Nanjing, China, 793.
540. Kabanov, V. A., Efendiyev, A. A., and Orudzhev, D. D. (1979). *Vysokomol. Soedin., Ser. A*, 31:589.
541. Efendiyev, A. A., Orudzhev, D. D., and Kabanov, V. A. (1980). *Dokl. Akad. Nauk SSSR*, 255:1393.
542. Nishide, H. and Tsuchida, E. (1976). *Makromol. Chem.*, 177:2295.
543. Nishide, H., Deguchi, J., and Tsuchida, E. (1977). *J. Polym. Sci., Polym. Chem. Ed.*, 15:3023.
544. Gupta, S.N. and Neckers, D. C. (1982). *J. Polym. Sci., Polym. Chem. Ed.*, 20:1609.
545. Braun, U. and Kuchen, W. (1984). *Chem. Ztg.*, 107:255.
546. Leonhardt, A. and Mosbach, K. (1987). *React. Polym.*, 6:285.
547. Japanese Patent (1969) 486,190.
548. Japanese Patent (1977) 5,719,209.
549. Japanese Patent (1983) 5,827,782.
550. U.S. Patent (1979) 4,173,555.
551. U.S.S.R. Patent (1972) 443,869.
552. Japanese Patent (1964) 498,048.
553. Pomogailo, A. D. and Klyuyev, M. V. (1985). *Izv. Akad. Nauk SSSR Khim.*, 1716.
554. Suvorov, A. L., Khonina, T. G., Kodess, M. I., and Podolsky, A. V. (1991). *Zh. Obshch. Khim.*, 61:1383.
555. Woell, J. B., Fergusson, S. B., and Alper, H. (1985). *J. Org. Chem.*, 50:2134.

INDEX

D

E

F

G

H

I

K

L

M

N

O

P

R

S

T

DATE DUE

GAYLORD PRINTED IN U.S.A.